Does It Matter?

Does It Matter?
Why Christians Should Care About the Young vs Old Earth Debate

Does It Matter? Why Christians Should Care About the Young vs Old Earth Debate
Copyright © 2018 Ronald C. Marks
All rights reserved. No part of this publication may be reproduced, stored in a retrieval system, or transmitted in any form by any means, electronic, mechanical, photocopy, recording, or otherwise, without the prior permission of the publisher, except as provided for by USA copyright law.
Cover design: Ronald C. Marks
First Printing 2019
Unless otherwise indicated, Scripture quotations are from the *ESV® Bible* (*The Holy Bible, English Standard Version®*), copyright ©2001 by Crossway, a publishing ministry of Good News Publishers. Used by permission. All rights reserved.
All emphases in Scripture quotations have been added by the author.
ISBN-13: 978-1-7323108-3-4 (Kindle e-book)
          978-1-7323108-4-1 (paperback)

Acknowledgments

As all writers with families, I am deeply indebted to mine. Particularly to my amazing wife, Joann for her patience and encouragement. Thank you, Joann, for the uncounted hours of assistance in intellectual, theological, and personal advice and counsel. Thank you both for your love and compassion.

To the many friends who have encouraged me along the way.

The Creation Study Group in Greenville, South Carolina, for providing the opportunity to give the presentation that lead to this book. Your tireless work for the Gospel has impacted many lives.

I am grateful to the work of God through so many others in the Body of Christ. I have seen far only by standing on the shoulders of giants.

Soli Deo Gloria

"Genesis 1 is just as true as Exodus 20, which gives us the Ten Commandments. It's just as true as Isaiah 53, which describes the suffering servant who would be the Messiah and bear our iniquities. It's just as true as Matthew chapter 1, which indicates that Jesus was to be born of Mary and to be the savior of the world. It's just as true as John chapter 3 which says you must be born again. It's just as true as any other and every other part of scripture."
John MacArthur, sermon on Genesis 1[1]

---

[1] Creation Day 6, Part 1, sermons Genesis 1:24–26 90-217 Jul 11, 1999

Table of Contents

ACKNOWLEDGMENTS	VII
CHAPTER 1 WHY CHRISTIANS SHOULD CARE ABOUT THE AGE OF THE EARTH	**1**
The Seatbelt Analogy	5
It Affects Our Essential Theology	8
Not All Stories are Created Equal	10
We Celebrate Real Events	12
CHAPTER 2 THE CRISIS OF AUTHORITY	**19**
Authority Affirmed: God Is	29
Authority Denied: God Is Not:	31
CHAPTER 3 THE CRISIS OF IDENTITY	**33**
Identity Affirmed: Made in the Image of God	36
Identity Denied: We Don't Know	42
Identity Denied: We Don't Know About Us	44
Identity Denied: We Don't Know About Family	51
Identity Denied: Gender No Longer Matches Gender	59
Identity Affirmed or Identity Denied	62
CHAPTER 4 THE CRISIS OF UNDERSTANDING	**69**
God's Word Affirmed: What we know has an anchor in Real Space and Real Time	83
God's Word Denied: Unable to Think Right	91
God's Word Denied: Preemptive Surrender	99
CHAPTER 5 THE CRISIS OF THE GOSPEL	**105**
God's Word Affirmed: Sin is Real, Knowable, Definable	105
Genesis of the Gospel Crisis	107
The Gospel Demands Clarity Regarding The Need	113
What are We Repenting From?	117
God's Word Affirmed: Identification of and Repentance from Sin	120
God's Word Denied: a Crisis of the Gospel	120
CHAPTER 6 WHAT MUST WE DO?	**123**
A Biblical Worldview	128
How Careful Should We Be With The Word of God?	130
ABOUT THE AUTHOR	**135**

CHAPTER 1
WHY CHRISTIANS SHOULD CARE ABOUT THE AGE OF THE EARTH

This book is *for* Christians. To encourage Christians in thinking properly about the age of the earth. I know that may sound strange – that Christians should need to be encouraged to believe what the Bible clearly states. However, in these later days, when Western Culture has rejected many of the Biblical truths that once defined a common worldview, many Christians find their thinking on many topics is no longer as clear as it once was.

Christians are finding more and more that we must draw the line on what we understand the Bible clearly states. Particularly on issues the culture has identified as offensive. When the culture characterizes Biblical truth as unacceptable, that same culture marks Christians who speak these truths in the same way. Unacceptable. Issues that include the definition of marriage, and proper human sexual relations, and the identification of gender, and the age of the earth.

If you are a Christian, you may not realize your beliefs about marriage, sex, gender, and other topics are affected by what you believe about whether the created universe is thousands of years old, or the evolved universe is billions of years old. Your church may have recently begun a specific ministry to homosexuals, and your pastor had a series of sermons on the need to reach out to this community. During that time, you and others have come to realize, based on the pastor's teaching, that homosexual people can be welcome in your church. You've had to change your mind about what may or may not be sin. Or, you've been thinking about and studying the impact of a predominantly white culture on Christian truth and become much more sensitive to your own inescapable sin of racism.

What you have not thought about is the age of the earth. To you, and to your friends, the age of the universe is a settled question that has been clearly established by science. Like many Christians, you either haven't thought about it, or haven't thought about it very much. Or, if they have, you have rejected the young-earth understanding and preferred the evolutionary story for how things came to be.

Here is the bad news. Tragically, those who choose to believe the universe is billions of years old have failed to comprehend the consequences of that belief.

If you are not a Christian, you may find portions of this book confusing. Or, though the thesis of this book addresses your culture and your place in it, you will find it a great challenge to appreciate the essential message. For those who are Christians, and those who reject the label, what you believe about the age of the earth matters.

Therefore, whether you are a Christian or not, let me encourage you to consider what you read with a determined openness. I am not asking for acceptance of what is presented without a critical analysis. Just the opposite. You are encouraged, if not outright exhorted, to consider this book with as much honest critical analysis as you can muster.

Let's begin by openly acknowledging that an effective critic is prepared for the illumination of analysis to occur in both directions. An honest intellectual has the humility and readiness of being prepared to consider that some notions or ideas they hold may not necessarily be correct. There is a proper and necessary distinction between what is true and what we believe to be true. Knowing there is a distinction is an important step in the pursuit of what is True.

Why is this book *for* Christians? Simply put, it is because we have been charged with dissemination of the Gospel. It is Christians who are tasked with faithfully and carefully sharing the Good News. This isn't something limited to an event, or tacked onto the end of a church service, or slyly slipped into a religious movie. It is the real Gospel that is an integral part of our worship that occurs with every breath, every heartbeat, every thought. That Gospel. The real one. It is too important to not proclaim the full and clear truth with clarity and boldness.

Here is where the distinction comes to play for the first time in this book. Although the creation vs. evolution debate may not seem essential to the presentation of the good news of God's regenerative work, it is essential to the good news of God's redemptive work. You do not have to believe the earth or universe is only 6,000 years old to come to a necessary understanding of salvation. An acknowledgment that God created the earth and universe out of nothing is not necessary to understand and repent of your sin and turn and confess the gift of God's forgiveness purchased through the death of Jesus Christ. How old the universe is does not impact a person's knowledge of their sin or of God's payment of the penalty for their sin. Belief in a 13.799±0.021 billion year-old universe[2] will not prohibit a person from total and complete acknowledgment of Jesus Christ as their Savior and Lord.

Yet, the presentation of the Gospel to the unsaved is based in understanding of being convicted under the law – recognition and acceptance of the substitutionary death of Jesus Christ – repentance and confession of Jesus as Savior and Lord. And, our growth under and in that same Gospel is founded on our understanding of who God is, what He is doing, and what He will do.

There is an important piece of knowing Whom the Savior is. For Christians, this impacts what you believe is true and True. Both words mean that something is correct and right. Uppercase "T" truth is used to distinguish ultimate truth that can only be revealed. Lower case "t" truth for any statement that simply matches what is correct and real. Having a proper ability to determine if what we believe is True or true affects our understanding of whom we serve. What we believe about the historical truth in the book of Genesis has an impact your ability to understand and properly impact the culture in which you live. It impacts the message we have been tasked to carry.

When young-earth Christians discuss the age of the earth with other Christians, they encounter an all-to-common attitude. This attitude is particularly prevalent with Pastors and church leaders. These Christians view the age of the earth and the consequences of holding a young-earth view as "just not important enough". For these Christians, in the bigger scheme

---

[2] The current accepted age of the universe as of the writing of this book in November, 2019. *Planck Collaboration* (2015). "Planck 2015 results. XIII. Cosmological parameters", *Astronomy & Astrophysics*, 594: A13, page 32, Table 4. arXiv:1502.01589

of all-things-church, this one isn't important. Pastors, elders, lay-leaders are all very busy and have professions, lives, ministries that are completely filled. Youth leaders who cannot add one more thought to the burden they are carrying. This topic of "young-earth or old" just isn't important enough for them to have given it much attention.

To them, it doesn't matter. At least, it doesn't matter enough. In their thinking, the age of the earth has no consequences in the daily pressures and their need to minister to the Body of Christ and to the culture.

The Seatbelt Analogy

Not caring about the consequences of your belief on the age of the earth does not make those consequences inconsequential. Just because you think the age of the earth does not matter does not mean it does not matter. This same thought process is often applied to seat-belt use.

Ignoring seatbelts, or claiming they aren't important to the operation of the car doesn't make them unimportant. Seatbelts serve a very important purpose. Most car occupants don't notice their importance until they encounter the event they were designed for. Accidents are a real potential component of the operation of a car. Just like airbags and brakes and crumple zones and collapsing steering wheels. Claiming seatbelt use is unimportant to driving or that they have no impact on how safe one drives doesn't make any subsequent crash less impactful. Claiming you are too busy with the important things required in driving a car does not make seatbelts unimportant.

The data clearly shows the importance of seatbelt use. Although this data is readily available, and often pushed in front of car users, people still choose to ignore the data and refuse to use seatbelts.

Figure 1 Head-on Collision. Seat-belt use is often neglected because the occupants think their use does not matter. Photo by Damnsoft 09 at Wikipedia commons is licensed under CC BY-SA 3.0. Photo converted to greyscale and edited to add color.

Here are the facts: Seatbelt use has been shown to reduce fatalities by 50 percent in head-on collisions, and by 74 percent in roll-over accidents.[3] Where I live in the state of South Carolina there were 500 traffic fatalities in 2019 in which seat belts were available to be used. Of those who died, 243 were not wearing seat belts.[4] That's 48.6 percent of the deaths that potentially may have been avoided by simply using a standard

---

[3] Kahane CJ. "Fatality reduction by safety belts for front-seat occupants of cars and light trucks: updated and expanded estimates based on 1986-99 FARS data". Washington, DC: US Department of Transportation, National Highway Traffic Safety Administration; 2000. Publication no. DOT-HS-809-199. Available at https://crashstats.nhtsa.dot.gov/Api/Public/ViewPublication/809199. Accessed March 3, 2018.

[4] "South Carolina Traffic Fatalities for 2018 vs 2019". South Carolina Department of Public Safety, data as of October 26, 2019. http://afc5102.scdps.gov/SCDPS_Exweb/SCDPS/HighwaySafety/TrafficFatalityCount_YTD

and available part of the car. In spite of this evidence, and state and local laws requiring seatbelt use, there are those who resolutely refuse to use them. Why?

Why where they not used? It is hard to understand why this readily available and simple to use device is ignored. Or, perhaps worse, purposefully not used. Particularly when the evidence so clearly shows using seatbelts saves lives. The answers, and there are many that are offered by those who refuse to wear seatbelts, can be distilled into one common cause. Those who have seatbelts available and refuse to use them do not (or did not) think it matters. They are seatbelt apathetic. Or, perhaps, seatbelt agnostic. The value of using a seatbelt while in a car is unimportant in the bigger picture of everything else in their world. Common excuses include, "It takes too long to put on", or "Seat belts are confining." They have convinced themselves the seat-belt will not make a difference. Even in the face of clear, unambiguous facts and experience that seatbelt use saves lives, they remain staunchly seatbelt un-involved.

Christians who conclude the age of the earth and the historical character of the book of Genesis do not matter are often participants to similar thinking. Pastors, elders, church leaders, as well as Christians in general, when asked about whether Genesis is historically accurate respond with an apathetic determinism. They genuinely think it does not matter. They are staunchly age-of-the-earth apathetic.

Ignoring this issue has not and does not make it unimportant.

It Affects Our Essential Theology

Many Christians have realized the theory of evolution is a theological worldview. Belief in evolution is a religious act. Evolution is founded on a claim of a non-supernatural created universe. It is a theory that, as Richard Dawkins states, made it possible for him to be "an intellectually fulfilled atheist."[5] Dr. Dawkins made this statement to note that atheists believe the theory of evolution gives atheists their own "creation story".

Most Christians are able to acknowledge the theological view of evolution is distinct and different from the theological view of Biblical creation. Atheistic evolution denies a supernatural source for the existence of the universe. Biblical creation confesses a supernatural God as the creator of everything. Atheistic evolution points to a pre-existing singularity as the source of a proposed "Big Bang". Biblical creation confesses a pre-existing God who has revealed Himself in His creation. Evolution is based in a theological statement – that there is no supernatural involvement in the origin of the universe. Evolution is not compatible with the truth claims found in Genesis.

Christian theology recognizes the title, claim, authority of the Messiah as an essential component of the identity of our Saving God, Jesus Christ, who is also our Creator God. His rightful claim as "King of the Jews" is linked to His rightful claim as "King of Kings".

This title of "King" is at the heart of the Gospel writers use of genealogies. In the Gospels of Matthew and of Luke, the lineage of Jesus has the purpose of establishing His claims of

---

[5] Richard Dawkins, *The Blind Watchmaker: Why the Evidence of Evolution Reveals Universe without Design*, W.W. Norton Co., 1988.

authority and identity. These genealogies are the same as those found in Genesis chapter 5.

This brings us to our first important point. If the genealogies in Genesis chapter 5 is only figurative or only an illustration, then we must also conclude the genealogies in Matthew and Luke are only figurative or illustrative, also. If they are not real and accurate in Genesis, they cannot be real nor accurate in the Gospels. If the persons and dates in Genesis 5 are not real, literal, events that occurred in real time and real space, we have no anchor for the genealogical linage recounted in the gospels of Matthew and Luke. If the genealogies of Genesis 5 only a figure or literary tool, they cannot be real in space and time. If they are not real in space and time, the Gospel claims of the Kingship of Jesus are not based in reality. They become false.

Old-earth creation must claim the genealogies are not records of real events from real history, just as evolution does. Young-earth creation confesses these are real events that are historically accurate.

This use of an anchor of real space and real time of Biblically-described events establishes a foundation for rational belief. Just as evolution is a theological claim, or based in a theological position, so is old-earth creation and young-earth creation. Old-earth creation insists the Genesis narrative of the creation events is figurative only. An old-earth creation view of Genesis limits the meaning to purpose and nothing else. To those holding this view, the book of Genesis shows us we have a purposeful God, but nothing more. There was purpose to the meaning, but not to any actual events.

In contrast, young-earth creation confesses the events of Genesis are real, having occurred in real space and a definable time. The events are filled with purpose and meaning in both their content and timing. However, that purpose is anchored in real, historical events. The sequence, the timing, the events are all real events occurring in real space with a corresponding definable historic time.

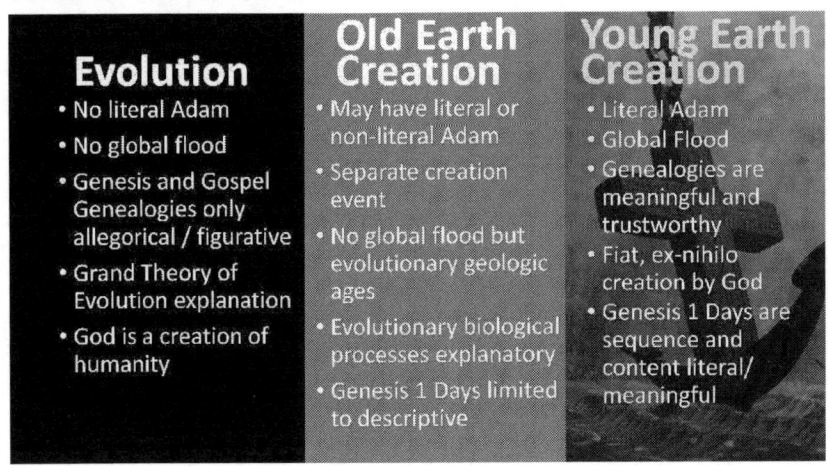

Evolution	Old Earth Creation	Young Earth Creation
• No literal Adam	• May have literal or non-literal Adam	• Literal Adam
• No global flood		• Global Flood
• Genesis and Gospel Genealogies only allegorical / figurative	• Separate creation event	• Genealogies are meaningful and trustworthy
• Grand Theory of Evolution explanation	• No global flood but evolutionary geologic ages	• Fiat, ex-nihilo creation by God
• God is a creation of humanity	• Evolutionary biological processes explanatory	• Genesis 1 Days are sequence and content literal/ meaningful
	• Genesis 1 Days limited to descriptive	

Figure 2 Comparison of the three worldviews of origins.

Not All Stories are Created Equal

If we can take historical narrative from the Bible and relabel it, or recategorize it, as nothing more than moral or beneficial stories, or in any other way identify or characterize the story as something other than real history, we open the door to use any story for establishing truth. Any story, regardless of source, can be used to not just transmit, but establish what we know to be ultimately true, and what we believe to be morally true. Why not use stories from other sources? As long as there is a claim, either external or internal, of containing and transmitting truth, the story would have the same authority as any other story. As long as the story has any hint of purpose that incorporates transmitting moral meaning and truth, it would be claimed as valid for that purpose. So, why not any story, whether they overtly claim to transmit ultimate truth? Could we use *Little*

House on the Prairie, or *Fahrenheit 451*, or *The War of the Worlds*, or *The Hobbit* to derive moral or any other ultimate truth? If the events in Genesis are not descriptions of real events that occurred in real space with a definable historic time link, then they are no different from any story, with the exception of claim in authorship.

That authorship claim includes a claim of authority beyond any human author. For the sake of this argument, though, we are viewing the authors as simply different.

The need for real events with an anchor in real space and real time is something we are familiar with. The modern mind has many examples. Let's use the example of prosecution of a crime to illustrate.

Evidence that can withstand cross-examination in a legal proceeding is set apart from evidence than cannot. The value of defensible evidence is recognized as useful in the context of establishing what is true. Consider the purpose of the lawyer in asking a witness, "Where were you when these events happened?" The lawyer is establishing the testimony in real space. "When was this?" is used to establish the testimony in real time and establish a time sequence for events. Placing events in a timeline acknowledges the timeline is real and important to the understanding of any truth derived from the events and related to the evidence. The attorney is building a case based on real events anchored evidentially in real space and real time.

We could also consider the process of good scientific research. Scientists are careful to document data collected during an experiment. And why not? It is real, valuable information from a real experiment. As the data is collected, a scientist is careful to document the place and time along with the observations as part of data collection. When reporting on this later, they have the real event to "reach back to" as having actually occurred in real space and real time. To defend the data as well as any conclusions derived from the experiment, the scientist will invoke the necessary real time the data was collected and note the real place the experiment was undertaken. Being able to place the data in real space with a definable historic time link is essential part of defending the trustworthiness of the data. It is also essential in supporting conclusions drawn from the evidence.

We Celebrate Real Events

We celebrate real events because they are real. They occurred in real space at a real time. This anchor in real space and real time is essential to the value of the celebration. And why celebrations are important. We celebrate birthday and wedding anniversaries because they remind us of a real event that occurred in real location at a known real time. Imagine a single person having a wedding anniversary. The very idea is ridiculous and confusing. In the United States, we celebrate the Fourth of July as representing the real day on which the Declaration of Independence was signed. For Americans, it represents the beginning of our nation. A real event anchored to a known date in history, July 4, 1776. We celebrate this real event because of its association with the real beginning of the United States based in an expression of a real document that begins with,

> *"When in the Course of human events it becomes necessary for one people to dissolve the political bands which have connected them with another and to assume among the powers of the earth, the separate and equal station to which the Laws of*

> *Nature and of Nature's God entitle them, a decent respect to the opinions of mankind requires that they should declare the causes which impel them to the separation."[6]*

The truths in God's Word are valid for many reasons. That they occurred in real space at a definable real time makes them tangible to us because we exist in space and time. Not accidentally. By the providential purpose of the Creator Who made space, and made time. So, those truths that we know from God's Word that are anchored in real space and time impact our lives for many reasons. One of those is because they are based in real events in real locations that occurred in real time. Jesus lived as a real human with a physical body. His body was not some metaphysical manifestation that simply appeared real. He was real. Christians have denied the wrong belief that His body was not real as heretical.[7]

Jesus became hungry and ate real food. His disciples watched Him get tired, sleep, and wakeup just like any other human does. His disciples as well as His enemies watched Him suffer and die. They witnessed and saw Him alive after he had died. Jesus had a real, physical body after dying a real, physical death. He had a real physical body after being raised from the dead, as the story of "doubting" Thomas reveals.

> *Then he said to Thomas, "Put your finger here, and see my hands; and put out your hand, and place it in my side. Do not disbelieve, but believe." Thomas answered him, "My Lord and my God!" Jesus said*

---

[6] The Declaration of Independence, In Congress, July 4, 1776.

[7] Docetism and Gnostic heresies taught Jesus' body was not real. Instead, these heresies teach Jesus body was an illusion of some type. This teaching was rejected as heretical. In the first letter of John (1 John), there is a clear statement about the nature of Christ that repudiates this dangerous belief. "By this you know the Spirit of God: every spirit that confesses that Jesus Christ has come in the flesh is from God," 1 John 4:2

to him, "Have you believed because you have seen me? Blessed are those who have not seen and yet have believed." John 20:27-29

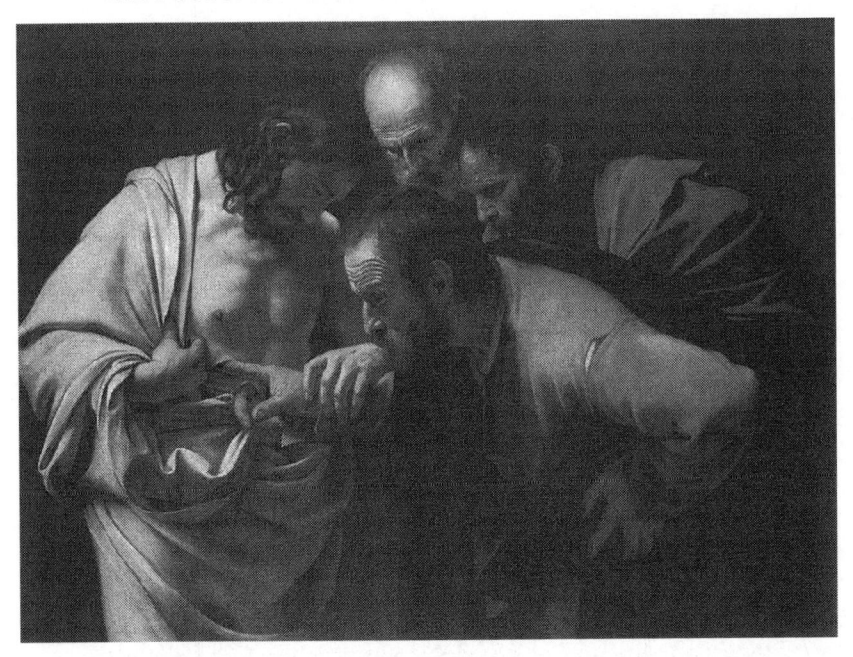

Figure 3 A depiction of real events with potentially a bit of artistic license. . I've always appreciated this depiction by Caravaggio as it represents the risen Savior unafraid of the intimacy of proving to Thomas that He was real. That his real body was proof of a real resurrection from the dead.[8]

We can place these events on a real timeline and in a real location. We know how long ago these events occurred and we can identify the places where they occurred. We visit and celebrate these locations because they are real locations associated with real, important events from which we are given ultimate truth.

The Bible is a real written account. It isn't just good stories that teach us important life lessons. C. S. Lewis' Space Trilogy and Narnia Novels, or J.R.R. Tolkien's Novels of Middle Earth all have great life lessons in them. But, they aren't real.

---

[8] Michelangelo Merisi da Caravaggio, *The Incredulity of Saint Thomas*, 1602, oil on canvas, Sanssouci Picture Gallery, Potsdam, Germany, public domain.

They are fantasy. If both Lewis' and Tolkien's novels and the Bible all contain great life lessons, does it matter if any of it is real? The answer is, "Of course". First, we know that Lewis and Tolkien and others have drawn on truth from God's Word in weaving their stories of fantasy. But, they are still fantasy. Without the location-time anchor, we could use any story. The Bible stands apart because it contains real, accurate history. The events did happen. And they happened at a time we can establish and in locations we can recognize.

A real physical David slew a real physical Goliath. A real physical Daniel was thrown into a real physical lion's den. And, there was a real, physical Adam and real, physical Eve that existed in a real physical place at a time that can be known.

The consequence of not wearing a seat belt can be devastating. Even though seat belts have been shown to matter in the event of an accident, many individuals still choose not to use them. The consequence of believing the old-earth vs young-earth debate doesn't matter has also been devastating. In the collision of Biblical ideas on a non-Biblical culture, poor thinking about what is in the first eleven chapters of Genesis has had damaging consequences. We see these in full effect in our culture.

Let's get to the heart of the matter. The culture needs to see and hear the Gospel. All of the Gospel and all the truth God has revealed. But, which parts? Which truth? What parts of God's Word should we leave out? Genesis is the foundation. Pastor John MacArthur makes it very clear.

> *"If Genesis 1-3 doesn't tell us the truth, why should we believe anything else in the Bible? Without a right understanding of our origin, we have no way to understand anything about our spiritual existence. We cannot know our*

> *purpose, and we cannot be certain of our destiny. After all, if God is not the Creator, then maybe He's not the Redeemer either. If we cannot believe the opening chapters of Scripture, how can we be certain of anything the Bible says?"*[9]

What you believe about the book of Genesis does matter. It matters what you believe about what it contains. It matters what you believe the book of Genesis is. Whether it contains events that occurred in real space and real time. How you characterize the book of Genesis affects everything else you believe. It impacts your theology and your worldview.

This impact on theology and worldview is seen in every classic arena a worldview addresses. How we think about the first eleven chapters of Genesis impacts how we think about four defining worldview questions:

1. Is there a supernatural? Or is the natural all that exists? Is there a god?
2. Who are we?
3. How do we know what we know? What is truth?
4. Is there anything beyond this life? Is there life after death?

---

[9] John MacArthur, "How Important Is Genesis 1-3?", web article, August 27, 2009, https://www.gty.org/library/articles/A176, accessed Feb 19, 2017.

Ultimate Truth in the Book of Genesis

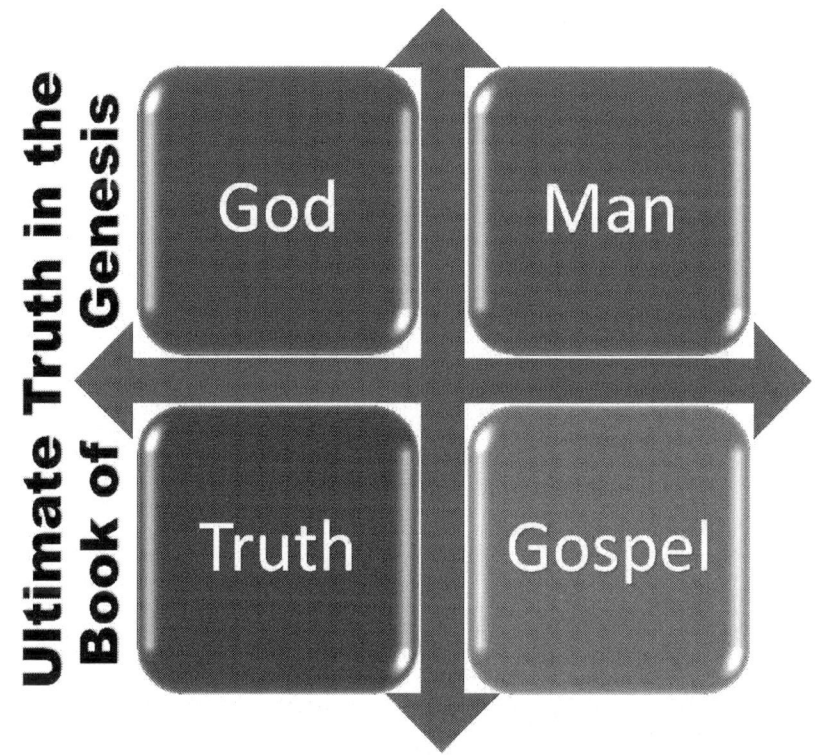

God	Man
Truth	Gospel

Figure 4 Worldview address four essential areas of ultimate questions.

Every one of these defining questions is answered in the first eleven chapters of Genesis. They are all answered in the creation and fall account described in the first three chapters. What you believe about Genesis is comprehensive in affecting what you believe:

1. About God
2. Man (mankind or humanity)
3. Truth (and truth
4. The Gospel.

A neglect of Genesis and subsequent rejection of a young earth perspective has resulted in wrong thinking about God with the consequences of a *crisis of authority*.

It has produced wrong thinking about mankind with the consequence of a *crisis of identity*.

It has affected how we think about Truth and produced a *crisis of understanding*.

And, it has corrupted how the Gospel is presented and lived out which results in a *crisis of salvation*.

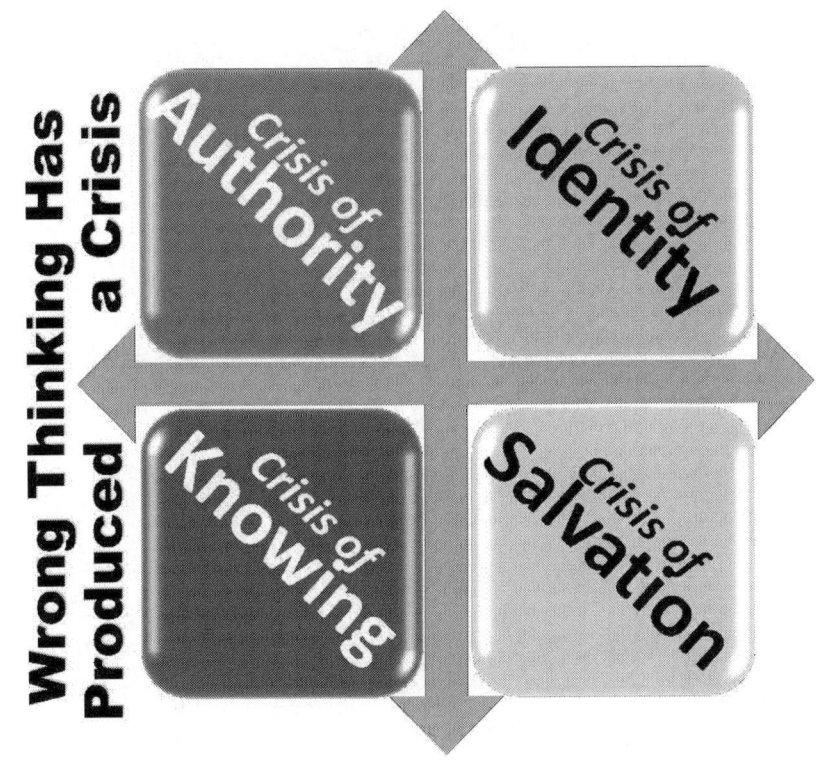

Figure 5 The crisis in essential areas of knowledge produced by wrong thinking.

Chapter 2
The Crisis of Authority

Modern culture has not only abandoned a Biblical view of authority, it is actively rejecting it. Many Christians, perhaps because of this constant exposure to the cultural rejection of authority, have also become confused regarding authority. Authority based in Biblical revelation. Christians must have a Biblical perspective of authority. That perspective is based in a confession that all authority derives from the ultimate authority, God. Authority exists because God, the ultimate authority, exists and has established authority and subordinacy. He alone sets up authorities.

> *Jesus answered him, "You would have no authority over me at all unless it had been given you from above. Therefore he who delivered me over to you has the greater sin." John 19:11*

Jesus clearly states that earthly authority is given by God - who is the ultimate authority.

And, we are to submit to proper authority in the home and in government.

Any culture without this foundation informed by or based in Biblical truth will be characterized by a rebellion and rejection against authority that is not contained within the individual. Proper authority that reflects Biblical patterns and purposes will be rejected in favor of self and individual authority.

> Let every person be subject to the governing authorities. For there is no authority except from God, and those that exist have been instituted by God. Romans 13:1

In submitting to authority we confess two things. First, that God is the ultimate authority and all authority derives from Him. Second, we have been called to submit to authority, which becomes our act of obedience and worship. Submission to authority is an act of worship. It confesses God is the ultimate authority.

Any culture without this foundation informed by or based in Biblical truth will be characterized by a rebellion and rejection against authority. They only authority recognized by that culture will be one that derives from the individual. Proper authority that reflects Biblical patterns and purposes will be rejected in favor of self and individual authority.

That is the current culture.

Many early church leaders warned about the consequences of this rebellion. Paul, in his Second Letter to Timothy, noted godlessness in the last days would be characterized by those who were "disobedient to their parents" (2 Timothy 3:2). These are individuals who reject the God-ordained authority of parents over their own children. Jude, writing to defend the Gospel and warning against the effects of apostacy, noted,

> *"Yet in like manner these people also, relying on their dreams, defile the flesh, reject authority, and blaspheme the glorious ones." Jude 1:8*

This crisis of authority is one of the defining characteristics of our generation. However, it didn't begin with us. It is as old, as, well frankly, dirt. Challenging authority, questioning authority, desiring to take authority away from the rightful source began long ago. In a garden. A perfectly created garden.

> *Now the serpent was more crafty than any other beast of the field that the LORD God had made. He said to the woman, "Did God actually say, 'You shall not eat of any tree in the garden'?" And the woman said to the serpent, "We may eat of the fruit of the trees in the garden, but God said, 'You shall not eat of the fruit of the tree that is in the midst of the garden, neither shall you touch it, lest you die.'" But the serpent said to the woman, "You will not surely die. For God knows that when you eat of it your eyes will be opened, and you will be like God, knowing good and evil." Genesis 3:1-5*

Do you see the subtly of this attack? God had given Adam and Eve clear instruction. He had taught them right from wrong. Eating the fruit didn't open their eyes to know what right and wrong were. They already knew. Eat from any tree in the garden except this tree equals being obedient to God – this is right. Disobey God and eat from this particular tree – this is wrong. God was the one who determined what was the right thing to do and which things (eating from that particular tree) were wrong. God was the authority in showing these first two individuals what was right and what was wrong. God has the authority to determine right behavior, and then reveal that to mankind.

Multiple lies were told that day. The first was a denial of the authority of God. When the serpent directly contradicted God. God said, "You may surely eat of every tree of the garden, but of the tree of the knowledge of good and evil you shall not eat, for in the day that you eat of it you shall surely die." (Genesis 2:17) The serpent challenged God's authority. He denied God's authority. He introduced confusion about authority with the result of producing a crisis of authority. Who is the authority? The serpent told Eve she could be, just like God was. The serpent said, "and you will be like God, knowing good and evil." (Genesis 3:5) Eve was told she could be the author of deciding what is good and what is evil. That she could have the authority to determine what was right and what was wrong. Just like God does. It was a lie. It is still a lie.

I've heard many people argue that this is only a story of how humans came to know right from wrong. That God only intended for this to be an analogy of the first rational humans, and how we became rational by disobeying God. If anything illustrates the truth contained in this real event, it is that humans have taken this real historic event and attempted to repackage it as non-history. In doing so, they claim the right of authorship. They claim the right of determining right from wrong. A crisis of authority.

This sin produces a particular fruit. A fruit that is at the heart of every sin. When we look to what God had said and begin to question if He really meant it. Or, when we begin to think we can change it to modernize, or correct, or refine His Words.

Once again, let me quote John MacArthur. His illustration of this crisis of authority using the same historical event from Genesis 3.

> *"Listen to this, for the first time, the most deadly spiritual force ever released was by that question released into the world...and it is this deadly force, the assumption that what God said is subject to our judgment. That's the issue, that's what launches this entire attack. 'Hey, Eve, let's talk about what God said and how we feel about it.' Really. Here is the deadliest force that has ever been released in the world and it's covertly smuggled into the world by means of Satan using a reptile as an instrument. The assumption is that what God says is subject to our judgment, our evaluation, our assessment. Now listen. All temptation, all temptation, I mean that, all temptation starts with the idea that we have a right to evaluate what God has said or required. It is subject to human judgment."*[10]

We see the rotting fruit of this crisis embedded in the fabric of our world today. Challenges to authority in our homes, churches, schools, workplace, government. We are experiencing the fruit of rejection of authority. There is rebellion against police officers as they work to protect us from criminal activity. Rebellion against laws that establish national boundaries and immigration.

The roots to this crisis of authority are deep within our culture and across many parts of the Western World.

Here's the question we need to ask, placed in the context of "Does It Matter what we believe regarding the age of the earth?" Our culture in the US was once a Christianized culture. Christianized, not Christian. It was heavily influenced by and operated on Truths from God's Word. These Truths could be seen

---

[10] John MacArthur, Sermon, "The Fall of Man, Part 1", 90-238, March 5, 2000, Grace To You. Transcript https://www.gty.org/library/sermons-library/90-238/the-fall-of-man-part-1, accessed March 5, 2018.

in the ideals and morals in operation between strangers meeting on the street (respect, kindness, tolerance, value) as well as in business transactions. Anyone who lived under the Christianized culture has observed this change. As the church paid less attention to the necessary value for authority and submission to authority, transmitting this important truth to the culture ceased.

The church lost its voice into the culture. We stopped speaking both ultimate truth and truth to the culture. We replaced speaking ultimate truth and confrontation of sin with overtures of peace and desires for friendship. In our attempts to contextualize and make the Gospel inoffensive, we lost sight of an essential truth. As we wanted the culture to want us, we laid aside the truth that the Gospel is offensive.

Western Culture has gone from Christianized to anti-Christian. The same change has happened, or is happening in the United States, but more slowly. In some areas, the culture is post-Christian. In other areas, the culture is better characterized as having come so far away from any impact by Christian truth and Christian ideas, that the culture is pre-Christian. And, in increasing areas, the culture is anti-Christian. The crisis of authority is both cause and effect of these changes. How did this happen in a nation that has so many Christian churches? How did this happen in our culture that was at one time infused with Christian ideals?

The answer lies at the feet of the church. Many churches set aside the authority of God's Word which was able to speak to every area of their lives and the culture around them. Instead, they began to elevate the claims of atheistic science, humanistic philosophy, or secularistic society.

The trustworthiness of Genesis as containing genuine historical narrative was one of the first areas compromise took hold. Not only regarding the creation, but also the global judgement described by the flood of Noah.

Denial of a global flood is a direct challenge to the authority of God's Word. The clear record of Genesis 7 contains details of a real physical event that can be placed in real time and a real location. God ordained that details of days, months, and years would be specifically recorded and preserved. The location was just as specific. That location being the entire surface of the earth.

> *And all flesh died that moved on the earth, birds, livestock, beasts, all swarming creatures that swarm on the earth, and all mankind. Everything on the dry land in whose nostrils was the breath of life died. He blotted out every living thing that was on the face of the ground, man and animals and creeping things and birds of the heavens. They were blotted out from the earth. Only Noah was left, and those who were with him in the ark.*
> *Genesis 7:21-23*[11]

The challenge to authority claims this wasn't a global flood. It couldn't be. A global flood is just too much. To this thinking, "all" doesn't mean "all". The list of what died is meaningless, regardless of what is stated in Genesis 7. "All mankind" cannot mean what "all mankind" normally means. "Blotted out every living thing" means something else. It has too according to those who are rebelling against the authority of God's Word.

If it has too, and what is written doesn't mean what it normally means, we have a *Crisis of Authority*.

When the postmodern mind claims truth is unknowable, it is simply another restatement of the modernist rejection of authority that is not contained in self. When God gave the commandments to the people of Israel, He demonstrated the source of all ultimate Truth. I am using the uppercase "T" to designate this type of Truth

---

[11] Emphasis added.

from all other things we call truth and know to be true. The existence of God is Truth that is provided by the self-revealing giver of knowledge of Himself. This type of Truth is different from the normal truth we express in statements like "The sky is blue" or "ice is cold". Ultimate Truth is revealed as it is given to us from God.

Elevating human reason over revealed Truth (ultimate Truth) is at the root of deciding to believe an evolutionary age of the earth (old-earth) over a Biblical age of the earth (young-earth). When we choose – and we are choosing – whether we hold to an old-earth or a young-earth view of creation, or we ignore the issue assuming it doesn't matter, we are making a statement concerning authority.

Clear thinking and honesty in our conclusions is essential to any ultimate issue answer. Just as it is with this one. To choose an age of the earth of 4.5 billion years is to acknowledge the evolutionary theory claims as authoritative. To choose an age of the earth of 6,000 years is to acknowledge the Bible, revealed truth, as authoritative.

Clarity and honesty also require an acknowledgment this is not an argument for anti-intellectualism or anti-rationalism. For a person to believe the earth is only 6,000 years old is neither anti-rational nor anti-intellectual.

Let me state that as strongly as I can. It is unbiblical to believe we must set aside intellect and "just believe". It is anti-Biblical to state we do not need rational arguments. God is both the author (source) of intellect and of reason. He calls us to use our ability to reason,to be rational.

> *"Come now, let us reason together, says the LORD: though your sins are like scarlet, they shall be as white as snow; though they are red like crimson, they shall become like wool." Isaiah 1:18*

When Jesus Christ was asked "What is the greatest command?", His response clearly showed the design of God was that our intellect is an essential and inseparable part of our worship of Him. He stated,

> *"You shall love the Lord your God with all your heart and with all your soul and with all your mind."*[12]

A claim that we must give up good thinking to believe God is to be guilty of a crisis of authority just like claiming we must set aside belief and trust our intellect alone.

When it comes to the age of the universe and the age of the earth, some will disagree, and claim we can hold to a 4.54 billion year age of the earth and still do so under an authoritative word of God. They invoke a hermeneutic of scripture which claims the times and events in Genesis 5 are incomplete. Or, in some way not to be used as literal times. This claim often invokes an argument that the ages of the patriarchs listed in Genesis 5 must in some way or another not be meant to represent real times. This hermeneutic requires that God incorporated these dates with some other purpose than to represent real, actual years. In their mind God was not clear. Therefore, we can look to human reason for clarity. This creates a crisis of authority as God's Word is made subject to man's need to define a different meaning.

Yet, when science also claims there is no resurrection from the dead, those who hold an old-earth hermeneutic shift their hermeneutic. When it comes to the resurrection of Christ, old-earth holders claim we can accept God's word as authoritative regarding the resurrection. Science, however, doesn't agree. Science doesn't turn to the old-earth believing Christian and say,

---

[12] Matthew 22:37, Mark 12:30, Luke 10:27

"Oh, OK. As long as you are now talking about resurrection from the dead, we'll agree that could happen."

No. There is a clear disagreement. Science also is clear that once a person dies, they are dead. People do not rise from the dead.

We are faced with deciding which to choose. Which to believe. Will we challenge the authority of God and claim the decision about what is true is up to us?

There are other Biblical, historical accounts of people dying and being raised back to life. The resurrection of Lazarus from the dead is one particularly important example (found in John 11:38-44, though the full story contained in John 11:1-44 is astounding). Science claims this does not, cannot, did not occur. Just like creation in six-days only 6,000 years ago. "Didn't happen", says science. We have a choice of making ourselves the authority for truth or confessing revealed truth for what it is. Just as did the people who witnessed the resurrection of Lazarus. Or, those who heard about it. This event made the religious leaders very angry, because it caused people to believe that Jesus was Who He said He was – God.

By the way – why did Jesus raise Lazarus from the dead? He didn't have to from any human standpoint. There wasn't some kind of Biblical law Jesus was upholding. In actuality, Jesus was superseding the physical, natural "laws" or more properly, physical natural processes. Why did He do this? In both John 11 (the account of the resurrection of Lazarus) and John 12 (the Pharisees planning to kill Lazarus because his resurrection was causing people to follow Jesus) show the purpose was to reveal Who Jesus Was. Jesus raised Lazarus from the dead so that we would believe Jesus was who he said He was.

Science also tells us there is no such thing as sin. Moral relativism teaches that right and wrong are constructs of culture and individuals. It can shift and change to whatever whim we desire. Because, after all, we are the highest authority. Science challenged and, in their minds, threw down the authority of God, and stepped into the place of the god they supposedly threw down.

However, the concept of sin requires acknowledgment of an external ultimate authority that has revealed what is necessary to not sin. Someone or something that exists outside of ourselves that is the source of knowing what is right and wrong.

Throughout the history of the church, Christians have created tools and systems to help learn and know orthodox theology. Some of the tools included catechisms, like C.H. Spurgeon's "A Puritan Catechism with Scripture Proofs". Or Confessions which were often a response to a heretical belief that had invaded Christian thinking. These tools help us understand doctrines that include the nature of sin. The New City Catechism identifies sin, as "rejecting or ignoring God in the world he created, rebelling against him by living without reference to him, not being or doing what he requires in his law—resulting in our death and the disintegration of all creation."[13] Rebellion against authority, or denial that an authority exists, doesn't make that source of authority cease to exist. It simply produces a crisis of authority in the rebellious creature.

Authority Affirmed: God Is

The very first verse of Genesis, which is, of course, the very first verse of the Bible, makes an incredible statement. Perhaps the most profound of all confessions that could be confessed. It is an introduction to all of scripture. It is a confession of a God who has

---

[13] *New City Catechism*, No. 16, "What is sin?", Copyright 2017 by The Gospel Coalition and Redeemer Presbyterian Church, published by Crossway, http://newcitycatechism.com/new-city-catechism/#16, accessed March 5, 2018.

no beginning. It makes a statement of the existence of God without making an argument or defense for the existence of God. God is simply stated as being.

> *In the beginning, God created*
> *the heavens and the earth.*
> *Genesis 1:1*

This is a confession of the existence of God. This is a confession of the pre-existing God. God Who existed before the creation. God Who has no start and no end. God Who is eternal. It is a rational statement of the existence of God, the transcendence of God, the eternal God. It is a proclamation that He is the First Cause.

This confession produces understanding that God is:

- Ultimate in power
- Ultimate in knowledge
- Ultimate in authority
- Clear in His actions

Does it matter when this confession is made?

Yes, since placing it here, at the beginning, as the opening statement, before the detailed description of the creation events and times is a statement of authority and importance. Prior to the details of creation, a confession of the character of God that is then revealed in the creative action and in the creation itself.

It was not placed here accidentally. God didn't select this opening statement because He needed a good opening to get things rolling. He does not need a hook to capture the reader's interest.

He placed this first. In sequence. At the beginning. Before time existed.

Before time existed. This statement establishes or fixes the start of measuring time. It places these events as having a real start. A point at which we can understand there is occurring a transition from things not being to things being. A fixed point in which we transition from not creating to creating. A fixed point that marks the transition from before time existed to the existence of time.

It is a real point. We may not be able, yet, just from this verse alone, to establish when this statement is specifically addressing. But, we can establish it references a transition and it confesses a cause. And not just any cause. The Cause. God the Creator.

All of these statements are true about God regardless of when the confession or proclamation is made. However, making them a priori communicates the authority for making every other statement.

Authority Denied: God Is Not:

> *For in some undetermined period of time, the LORD made heaven and earth, the sea, and all that is in them, and rested on the seventh undetermined period of time. Therefore the LORD blessed the Sabbath undetermined period of time and made it holy. Though, it doesn't really matter.*
>
> Exodus 20:11, *As Some Would Prefer It*

Challenging the authority of God and placing human reason as the source for what is correct and true has intended as well as unintended consequences. This challenge to authority is characterized by claims that Genesis chapters 1 - 11 do not contain historically accurate events. Because we are giving the authority to science or some other human mechanism, we consequently are making one or more of the following statements about God:

- "God limited Himself because ancient humanity couldn't understand what God actually did during the creation." This either intentionally or unintentionally claims that God was unable to communicate what He needed to, intended to, communicate. He was limited by early human's lower intellect.

- "Only with the advances in knowledge and modern science could we now finally understand that God took billions of years to create everything." Like the previous statement, this makes the claim of a limited God who had to wait until humans had developed sufficient intellectual understanding. Not God self-limiting for His purposes, but unable to overcome the limitations of His creation. Both this and the preceding statement are forms of chronological snobbery.
- "Genesis isn't about anything important other than to say that God created the universe. Reading anything else than that simple message from Genesis 1-11 is to go beyond what God intended." God as authoritative and able in communication requires that His Word is sufficient for all times. Even the early. Even the late.

Why is science correct regarding the age of the universe, yet not correct regarding resurrection from the dead? Who's right about sin? Is it good to submit to authority, or should we reject all forms of authority?

We are deep in a crisis of authority.

CHAPTER 3
THE CRISIS OF IDENTITY

In addition to causing a crisis of authority, ignoring or minimizing the historical, literal reality of the Genesis account produced a crisis of identity. It corrupts our understanding of who we are. If the events described in the creation account did not occur in real space as described, at a real time that could be known, the consequences are a crisis in mankind's identity. We lose the clear teaching of what it is to be *imago Dei*, an image bearer of God.

Genesis contains a clear, unambiguous statement of the creation of man. A creation that is distinct from the creation of animals. According to that creation difference, we are not evolved or advanced animals. We are different and distinct from them. The Apostle Paul emphasized this in his letter to the church at Corinth. Paul was encouraging those Christians concerning the resurrection and used the distinction of mankind from animal kind to illustrate the difference between the mortal and immortal.

> *For not all flesh is the same, but there is one kind for humans, another for animals, another for birds, and another for fish. 1 Corinthians 15:39*

The Genesis account is very clear that mankind's creation was special. The account uses new words and new actions not seen in any prior creative act.

> *Then God said, "Let us make man in our image, after our likeness. And let them have dominion over the fish of the sea and over the birds of the heavens and over the livestock and over all the earth and over every creeping thing that creeps on the earth." So God created man in his own image, in the image of God he created him; male and female he created them. Genesis 1:26-27*

In every other created event, the creation occurred by God speaking the object into being. He commanded, creation occurred.

> *And God said, "Let there be light," and there was light. Genesis 1:3*

> *And God said, "Let there be an expanse in the midst of the waters, ... And it was so. Genesis 1:6-7*

> *And God said, "Let the earth sprout vegetation, ... And it was so. Genesis 1:11*

> *And God said, "Let there be lights in the expanse of the heavens ... And it was so. Genesis 1:14-15*

> *And God said, "Let the earth bring forth living creatures according to their kinds ... And it was so. Genesis 1:24*

But, when God created mankind, a different method of the creative act is indicated. In creating man God becomes intimately involved in the creation in two ways. First, God proclaims intention to create man after His own image. Mankind, unlike any other part of creation, is to have a unique role as particular image bearers of God. That is a special, unique created purpose. It rules out an evolutionary source for our existence. The second

The Crisis of Identity

difference is in the method God used. In creating man, God formed us from the dust of the ground.

> then the LORD God formed the man of dust from the ground and breathed into his nostrils the breath of life, and the man became a living creature. Genesis 2:7

Formed us. The image produced by use of the word *formed* is that of a potter using his hands to squeeze and mold clay into the *form* he intends. The animals were called into existence. The first human was *formed*. After intimately forming the first human, God then breathed into the nostrils of the newly formed body. No other created thing had God involved in their creation like this.

These are real events. Not imaginative, or figurative, or allegorical, or illustrative. This really happened. The creation of man describes events that really happened in real space at a definable time. A real space – on the surface of the earth in a singular location. In real time, the sixth day of creation. A specific day selected by a purposeful Creator.

The old-earth creation origin story of human beginnings requires a separate set of events for the creation of the first humans. It incorporates some form of evolution and uses non-human ancestors to produce pre-modern humans. These pre-modern humans then produced, in some way that may have required a second miraculous intervention of God, the first modern humans. The first male and female human pair or group.

Theistic evolution, on the other hand, simply places humans completely within the evolutionary process. Yet, they too must account for the first humans. Whether theistic evolution or progressive creation story, the first bipedal humanoids are identified as those who had all the traits of intellect and moral reasoning combined with self-aware identity necessary for understanding of sin and personal responsibility.

Any old-earth or evolution-dependent origin of humans incorporates a process of undirected, random events that resulted in increase in complexity and specialization of being. Since these processes are identical for anything that exists, then the value of humans and being human has no basis. There is no special distinctiveness of being human. We cannot claim human life is more important than the life of a dog or banana tree without basing it in some type of difference. If we argue from complexity and self-awareness, the psychologist and philosopher will note these are derived from inside the person. They are internally sourced. We are calling ourselves valuable because we decide we are valuable. But, then we can also decide that one type of human is more valuable than another type of human. And that becomes, and became, a justification (in our minds, at least) for all types of sinful thoughts and behavior.

Identity Affirmed: Made in the Image of God

God purposefully and intentionally created mankind. We are not the product of random processes or chance mutation. When we confess what is clearly and intentionally stated in Genesis chapters 1 through 3 regarding the creation of mankind as real, historical events, we can also confess what is derived from these events.

On May 18, 1980 in Skamania County Washington State, the United States of America, an event occurred that impacted a great many people. On that day, Mount St. Helens, which had been dormant for many years, became explosively active. That volcanic event is the cause of effects still observable today. Trees that were

The Crisis of Identity

Figure 6 A false color image of Mt. St Helens from the Landsat satellite survey in 1979. Vegetation is colored as red. The light area is the mountain with elevation "above the tree zone" and covered by snow. (NASA Earth Observatory web site. Public domain)

snapped off near the ground by the force of the eruption, a crater from the explosion, and other effects are still observable decades following the eruption. We understand the broken trees were produced by the blast that occurred in real space and real time. Comparing images from satellites shows a distinct difference. The story that describes the events to cause these changes is linked to the real event that occurred on May 18, 1980.

When Jesus was asked about marriage and divorce, His response used the real events of the creation of man to provide the anchor for meaning in His answer. He linked the truth of His day with the truth of the creation event that produced the foundation for correct understanding about marriage.

> He answered, "Have you not read that he who created them from the beginning made them male and female, and said, 'Therefore a man shall leave his father and his mother and hold fast to his wife,

Figure 7 Landsat image following the eruption of Mt. St Helens in 1980. Note the clear formation of the crater from the explosion. (NASA Earth Observatory web site. Public domain).

Figure 8 Landsat image of Mt. St Helens from 2016 clearly shows changes to the land which caused by the eruption 36 years prior. (NASA Earth Observatory web site. Public domain)

and the two shall become one flesh'? So they are no longer two but one flesh. What therefore God

The Crisis of Identity

has joined together, let not man separate."
Matthew 19:4-6

Jesus, Who is the Creator (He was there - therefore He knows what happened at the beginning), used this real event to answer the crisis of identity the Pharisees had regarding God's design for marriage. A crisis we have faced in modern western culture. And, as a culture, have failed.

Before we consider the impact of the answer in answering this crisis of identity, pause to notice how Jesus began His answer. He said, "Have you not read…" It is a question asked not for the sake of meeting a deficiency in Jesus' knowledge, but to provoke understanding in a deficiency of those who brought the challenge. This is a rebuke of pride in those who considered themselves the learned leaders of the nation of Israel. A rebuke that points unambiguously back to the Genesis account of creation as real, historical events. They had the clear, unambiguous, obtainable answer readily available to them. Yet, their question showed they had either rejected or neglected this clear teaching.

Just as those theologians were confronted by the words of the Creator, so are theologians today who deny the authentic historicity of the Genesis account. The same Savior modern theologians look to for the foundation of their faith has claimed that men and women were created as a male and a female at the beginning of creation. Not as proto-humans. Not as a group of pre-modern hominids who then became the first humans.

Our identity as image bearers of God is established in a real creation event at the beginning of time. It is founded in our creation as a male and a female as the creation account records.

> *So God created man in his own image, in the image of God he created him; male and female he created them. Genesis 1:27*

James, in his letter that became part of the New Testament, calls Christians into practical holiness by reminding them that we are image bearers of God. By looking back at a real event that occurred in real space at a real time, James exhorts Christians to control their actions and treat all people for what they are: image bearers of the Creator. He does so by contrasting humans with animals.

> *And the tongue is a fire, a world of unrighteousness. The tongue is set among our members, staining the whole body, setting on fire the entire course of life, and set on fire by hell. For every kind of beast and bird, of reptile and sea creature, can be tamed and has been tamed by mankind, but no human being can tame the tongue. It is a restless evil, full of deadly poison. With it we bless our Lord and Father, and with it we curse people who are made in the likeness of God. From the same mouth come blessing and cursing. My brothers, these things ought not to be so. James 3:6-10*

This knowledge affects how we treat one another, care for one another, value one another. A proper view of mankind, a proper anthropology, is based in a real creation of a real man and a real woman on a real day. On the sixth day of creation.

This *imago Dei*, this image of God, is more than a statement to make us feel "special". It is an essential component of our identity and our proper duty. As our identity, it must be something anyone can see. We must "wear it" honestly, proudly, humbly. It is not something we "carry" or "put on". It is something we already *are*. It cannot be taken off, set down, covered up. It is always there. It must be properly confessed.

To make a proper confession of being created in the image of God, we must confess we are created. Not evolved. We are purposefully and distinctly created differently from the rest of creation. Therefore, humans are not animals.

The taxonomic (or arrangement method) system developed by Carl Linnaeus places humans in the animal kingdom for purposes of classification. In his original classification of three Kingdoms, animal, vegetable, and mineral, humans are neither vegetable or mineral. Using the limitation of only those three choices, the assignment is correct. But, it is only correct because the system has a flaw in the classification scheme. Even though the biologic taxonomy places humans in the same kingdom as animals, we must confess our *imago Dei*. God did not place His image in anything else He created. Not in any animal. Only in man and woman. To be consistent with what is found in Genesis and confirmed (or emphasized) in the New Testament, Christians must distinguish animals from humans. We must apply this confession when we think, act, teach our families, and speak to the culture.

This image confesses we are created as distinctly male and distinctly female. Two clearly defined genders made to glorify God in the purpose He created in this distinction of gender. We must also glorify Him by celebrating and living the purpose of being male and being female. God designed and then placed one male and one female in a family. A covenant relationship in which the family is established and designed to flourish. A designed environment in which children are both produced and raised with the purpose of honoring God.

The *imago Dei* confession understands all of humanity as belonging to the same race. One race of humans. Skin color does not and should not be used to divide, segregate, or prioritize.

Identity Denied: We Don't Know

The crisis of identity dominating our current culture means we no longer identify as specially created humans. We are unable to properly confess we are the created image bearers of God. We no longer view humans as a special and particular creation designed to give God glory. This identity crisis has produced a belief that gender (or sex) is fluid, not something created by God for a specific purpose.

So, even though we cannot take off or put away this image that we are, we can deny it and act in ways that attempt to cover up or corrupt the image.

On June 26, 2015, after a long battle through the court systems that included multiple related cases in several US states, the issue of same sex marriage was addressed by the US Supreme Court. Same-sex couples demanded to have the same rights under the law with respect to marriage. These homosexual couples were asking for marriage to be radically redefined. In its decision, the Court made constitutionally protected right to marriage to include any combination of gender. Justice Anthony Kennedy, author of the majority opinion, wrote, "The Fourteenth Amendment requires a State to license a marriage between two people of the same sex and to recognize a marriage between two people of the same sex when their marriage was lawfully licensed and performed out-of-State." ([Obergefell v Hodges, 576 U.S. ___, 14-556 (2015), p. 28)

Justice Kennedy's opinion reflected the accepted confusion prevalent in a culture that has denied God's purpose in creating male and female. A confusion that emphasizes human reason and human feelings as the source of truth and understanding, and the foundation of ultimate truth. This confusion confirmed that the culture has lost the ability to read the message God placed in the revelation of creating male and female, and placing them in a life-

The Crisis of Identity

long covenantal union. In denial of this identity, the Obergefell vs. Hodges ruling anchors marriage in human ideals and emotion instead of a created and purposeful image that has the message of glorifying God. The decisions corrupted the understanding of Biblical love, dethroning God as the authority for defining right and wrong, and replaced Him with human feelings and desires.

> *No union is more profound than marriage, for it embodies the highest ideals of love, fidelity, devotion, sacrifice, and family. In forming a marital union, two people become something greater than once they were. As some of the petitioners in these cases demonstrate, marriage embodies a love that may endure even past death. It would misunderstand these men and women to say they disrespect the idea of marriage. Their plea is that they do respect it, respect it so deeply that they seek to find its fulfillment for themselves. Their hope is not to be condemned to live in loneliness, excluded from one of civilization's oldest institutions. They ask for equal dignity in the eyes of the law. The Constitution grants them that right.[14]*

God's intention of self-revelation in the created relationship of one man covenantally bound to one woman for life has been muddled to the point of unrecognizable. When Christians treat the Genesis account as less than real, historical events, they open the door for redefinition of marriage away from the message-bearing image established by God. When Christians think that it doesn't matter if the events in Genesis occurred in real space and real time, they give over the authority contained in the historical establishment of our identity. We have a crisis of no longer

---

[14] *Obergefell v Hodges*, 576 U.S. ___, 14-556 (2015), p. 28

knowing who we are. We have lost our real identity. And, we have lost the understanding that we have no way to regain it apart from repentance and self-denial.

Identity Denied: We Don't Know About Us

Scripture contains a clear, unambiguous statement of the creation of man. Mankind is not derived from or evolved from other animal life. Our creation is distinct and separate from the animals. It is different from the animals. It is special.

When describing the creation of man, there are different words employed and new actions described that are not used with any other creation event. In other creation events, God speaks apart from any other action.

> *And God said, "Let the earth bring forth living creatures according to their kinds—livestock and creeping things and beasts of the earth according to their kinds." And it was so. And God made the beasts of the earth according to their kinds and the livestock according to their kinds, and everything that creeps on the ground according to its kind. And God saw that it was good.*
> *Genesis 1:24-25*

Yet, in the creation of man, God's actions take on a very intimate presence and intention. There are two distinct characters of the creation of humans that make us distinct from the creation of animals. The first is in the inclusion of God's image in the creation of man.

> *Then God said, "Let us make man in our image, after our likeness. And let them have dominion over the fish of the sea and over the birds of the heavens and over the livestock and over all the*

The Crisis of Identity

> *earth and over every creeping thing that creeps on the earth." So God created man in his own image, in the image of God he created him; male and female he created them. Genesis 1:26-27*

God defines this creation of man as a creation after His own image. Of all the creation events, this particular, special event incorporated the image of God. God purposed for humans to be formed in His image. God intended for humans to bear His image. We are image bearers of God. That is a special, unique created purpose. It stands in stark contradiction to the evolutionary claim of unguided, unintentional, unmeaningful cause of our existence.

The second unique aspect of the creation of man is in the words used. Unlike the creation of animals (or stars, or the ocean, or anything else) which were simply "spoken" into existence, God became intimately involved with the creation of man.

> *then the LORD God formed the man of dust from the ground and breathed into his nostrils the breath of life, and the man became a living creature. Genesis 2:7*

The details included in the creation of man are breath-taking. Without any additional digging into the meaning, simply realizing what is being stated about our Creator in His care for this event is astounding. Unless, of course, it is only allegory, or metaphor, or some other literary tool instead of real space, real time, historical narrative. Then this story is only mildly interesting.

But it is real. It did occur in a real, definable space at a specific, definable time. Mankind was created on the sixth day of creation, after the animals had been created. This event took place in a real, clearly definable space. We may no longer know the exact location where this original garden was located, but it did exist. It existed on the surface of the planet we still occupy. Even though a catastrophic judgement destroyed the original surface, it is still the surface of the earth. A real place that exists in real time.

It was a real Creator who created a real, specific person. Although it is important to confess that the triune and pre-incarnate God does not have a physical body, He does not have arms or hands or fingers, this action is described by shaping, molding, pressing. The account describes God forming the man. The Hebrew word used for "formed" is the word יָצַר (yaw-tsar), meaning "to squeeze" or "press" into shape.[15] It is the same word that would be used to describe the action of a sculptor working with a lump of clay, molding and shaping until that ambiguous meaningless blob takes on definition, form, and identity.

The word use is powerful in giving clarity, not confusion, to what God did. Although God doesn't have "hands" in a way we would say a person has hands, and God is not limited by the creation as the created is. God entered into His creation and manipulated the material present there, forming it into the human shape. Forming the molecules at the atomic level into the shape of the DNA, the proteins, fats, collagen, hydroxyapatite, sugars, glycosamino-glycans, and all the physical parts of the body. Bending, attaching, connecting, relating all the interdependent pieces. Not by command. The record of the event clearly states God was more personally involved with this creative event.

Following this very personal act, God breathed into the newly formed body of man. The Creator becomes even closer to His

---

[15] *yâtsar (yaw-tsar')*, probably identical with יָצַר (through the squeezing into shape); to mould into a form; especially as a potter; figuratively to determine (that is, form a resolution): - X earthen, fashion, form, frame, make (-r), potter, purpose. Dictionaries of Hebrew and Greek Words taken from Strong's Exhaustive Concordance by James Strong, S.T.D., LL.D., Published in 1890; public domain.

created. The Hebrew word נפח *(naw-fakh')*[16] would remind us of the action taken in resuscitating a person who had stopped breathing. We "blow" into them a breath. Air that was in our lungs is passed into the lungs of the other person. This is an extremely intimate event. The two persons share their breath.

Here is a description designed to help us understand something we cannot otherwise. Whatever the breath of God is (and I do believe we can know what it is), He chose to use that part of Him to pass life into the lifeless body. That breath of God is purposefully different and unique in the creation of man.

None of the other parts of creation, including animals, received this "breath of God". The creation of man was something clearly different. Clearly set apart.

In summary, humans are a special creation in the greater creation with the distinct duty of being image bearers of God.

Old-earth creation acknowledges the necessary purpose and meaning of being created by God. This old-earth view must have an account for the creation of the first humans. The first male and female human pair. Yet, the old-earth view requires this first human pair or population to be undefinable in space and time.

A common story for this initial human presence incorporates some level of evolution and non-human ancestors.

Old-earth creation proponents will often acknowledge the necessary purpose and meaning of a special creation for humans. Therefore, they have to concoct a story that can account for the creation of the first humans within their evolutionary framework. They need some way to produce the first male and female human pair in a lineage of pre-human ancestors. Yet, the old-earth view

---

[16] Strong's: A primitive root; to puff, in various applications (literally, to inflate, blow hard, scatter, kindle, expire; figuratively, to disesteem): - blow, breath, give up, cause to lose [life], seething, snuff.

requires this first human pair or population to be undefinable in space and time.

One often used explanation for the first male and female pair proposes an unspecified, unknowable, speculative event in which humans either obtained or were made "fully modern humans". It is a story fabricated in an attempt to build a picture describing something that might have existed in some undefinable point of time and in some general but unknown location. There are no specific individuals. No real persons. Nothing is known about these first humans other than a concrete claim they exist in this paradigm. Because, they have to.

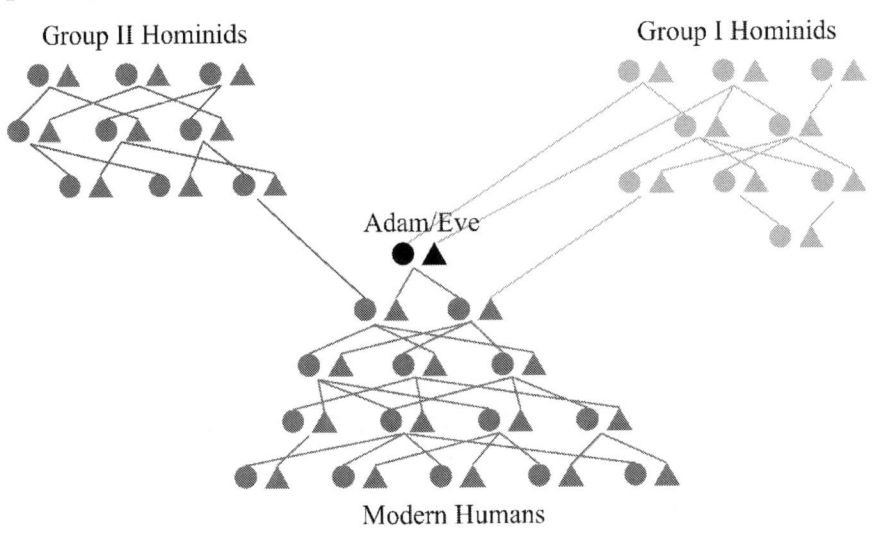

Figure 9 Old-earth Creation model for appearance of the first humans from pre-human hominids. This model allows old-earth creationists to incorporate evolutionary theory with scripture.

And old-earth creationists who build pictures like the one shown in Figure 8 will staunchly claim that just like young-earth creationists, they also believe the Bible and believe Genesis. One group, Reason to Believe, has the following statement on their web site regarding their position on the Bible.

The Crisis of Identity

> *We believe the Bible (the 66 books of the Old and New Testaments) is the Word of God, written. As a "God-breathed" revelation, it is thus verbally inspired and completely without error (historically, scientifically, morally, and spiritually) in its original writings. While God the Holy Spirit supernaturally superintended the writing of the Bible, that writing nevertheless reflects the words and literary styles of its individual human authors. Scripture reveals the being, nature, and character of God, the nature of God's creation, and especially His will for the salvation of human beings through Jesus Christ. The Bible is therefore our supreme and final authority in all matters that it addresses."*[17]

Although they use phrases that seem orthodox, such as "supernaturally superintended" and "God-breathed", they must not mean the same as it does to other theologians.

This old-earth belief does not include understanding the historical narrative as real events that occurred in a specific and knowable time and in a specific space. Even though some old-earth creationists claims the Bible as supreme authority, to include authority in areas of science, they give priority to science when there is a contradiction between scientific assignment of the age of the earth and the genealogical evidence in Genesis. For these old-earth creationists, there is no historical anchor of real space and real time that impactfully transmits the meaning and understanding of being an image bearer of the Creator. All old-earth models incorporate evolutionary processes that produce another "something" at "some unknown time". It just happened. Or, had to. Because, after all, here we are.

---

[17] Reasons to Believe, *Our Mission & Beliefs*, http://www.reasons.org/about/mission-beliefs, accessed June 16, 2018.

Progressive creation is an old-earth creationist model that attempts to merge the Genesis historical account with evolution by adding an extra-Biblical event. In this model, Adam and Eve occur in an already ongoing evolution of hominids. God decided to intervene in the already ongoing evolutionary processes and put His image into a hominid male-female pair, thus establishing the modern humans.

For these old-earth creationists, Jesus must have misspoken or was overly generalizing when He said,

> *But from the beginning of creation, 'God made them male and female.' Mark 10:6 (also Matthew 19:4)*

The Creator, the Savior, God incarnate said that humans were created male and female from the beginning. If the beginning doesn't mean the beginning, as required by any incorporation of evolutionary theory, then we are left with the task of finding out what God meant. And, since this the beginning wasn't really at the beginning, then perhaps none of the other events that occurred in some time sequence have any meaning either. God becomes the author of confusion and irrationality.

If at any point of human creation we are just the product of undirected, random processes, then the value of humans and being an image bearer of God has no basis. There is nothing clearly distinct about God placing his image in humans. We cannot claim human life is more important that the life of a dog or banana tree without basing it in some type of real difference. If we argue from complexity and self-awareness, the psychologist and philosopher will note these are derived from inside us. We are calling ourselves valuable because we decide we are valuable. But, then we can also decide that one type of human is more valuable than another. And then implement some "final solution" to rid the world of the other type. Or, enslave humans because they are a "lesser" or "less advanced" type. Both Hitler's final solution as well as race-based slavery were a result of this crisis of identity based in a confused anthropology.

Identity Denied: We Don't Know About Family

Why do we, as Christians, value a covenantal relationship between one man and one woman? Is there a spiritual value and benefit in marriage?

Have you ever pondered or wondered why God created two distinct sexes?

> *So God created man in his own image, in the image of God he created him; male and female he created them. Genesis 1:27*

And ordained they would live together, with purpose, for their entire lives? From the very beginning of creation, God created and established that a man and a woman would live their lives as *one*.

> *Then the man said, "This at last is bone of my bones and flesh of my flesh; she shall be called Woman, because she was taken out of Man." Therefore a man shall leave his father and his mother and hold fast to his wife, and they shall become one flesh. Genesis 2:23-24*

Christians have valued this union and communicated the importance of marriage to the culture around them for most of the church age. In doing so, the church glorified God by confessing His gift of marriage and acknowledging the purpose He placed in marriage. In every culture they were in, Christians modeled this truth of God's design for one man and one woman living in covenantal union for their entire lives. In this union, children were produced, trained, cared for, and raised to adulthood in which they became contributors to the culture.

Philosophers from ancient times recognized the value of a family unit. Plato, Socrates, and Aristotle all made reference to the relationship of family and government. Aristotle, for example, is known for teaching that cities were a natural outcome of human effort because it reflected community, which was related to having a family. The "seed" of any city-state is the family. Likewise, modern social commentators, at least until recently, have emphasized the link between strong families and a strong culture. William Bennett, known for his advocacy of moral self-reliance, defended a particular political party's actions by noting,

> *Republicans are talking about family values because they see the family breaking down in front of them. This talk isn't just happy, wishful thinking; this isn't nostalgia; this is reality. If we have stronger families we will have stronger schools, stronger churches, and stronger communities with less poverty and less crime. The family is the linchpin of society, both economically and socially. One of the best things Rick Santorum said during the primary, addressed to Republicans in particular, was that if you are serious about having smaller, more effective government, then you had better work at getting stronger families.*[18]

All culture benefits from long-term, stable families. When these stable families, as mini-societies forming the larger society, begin

---

[18] William Bennett, "Stronger Families, Stronger Societies", *The New York Times*, April 24, 2012, accessed online
https://www.nytimes.com/roomfordebate/2012/04/24/are-family-values-outdated/stronger-families-stronger-societies, June 18, 2018

The Crisis of Identity

to fail and disappear from the scene, the larger society suffers and eventually fails.

It is also true that throughout human history, many cultures became characterized by polygamous marriage and rejection of marriage (singleness) before those cultures crumbled. Both of these states, singleness and polygamy, grow out of a common affliction. That affliction is the pursuit of personal, individual, selfish wealth. Solomon gathered many wives and concubines, contrary to the instruction from God.[47] He did so to grow his political capital or wealth. Isaac Newton eschewed marriage, desiring to not be distracted from the "wealth" of gathering knowledge. He wanted the freedom to pursue what he valued most. When we no longer understand the value of personal sacrifice for the sake of healthy marriages and thriving families, our desire for having a family goes away. As our culture has made self-fulfillment the primary goal of life, the sacrifices and self-denial which are part of making families are antithetical to the culture's view of worth. The unintentional, or, it may actually be intentional, consequence is a culture with no strong families at its core. Even though these families are essential to the culture's function and existence, the culture will reject families. Our world is rejecting marriage as an expected or normative state. Individuals no longer value a covenantal relationship between one man and one woman.

For Christians who understand the value and importance of marriage, current trends should be disturbing.

The Pew Research Center analyzes and reports on the status of social trends using United States Census data and the American Community Survey data. A recent report shows that one-fourth of American young adults may never get married. Although a portion of these are cohabitating or may have short term relationships, the value and identity of a covenantal relationship is not present for these young adults. If you look around many churches, you'll see a reflection of these numbers. Young men and women who are attending church, who claim to be Christian, who even value Biblical truth, are choosing to remain single.

One-in-Four of Today's Young Adults May Never Marry

% never married, by cohort (at ages 25-34, 35-44, 45-54)

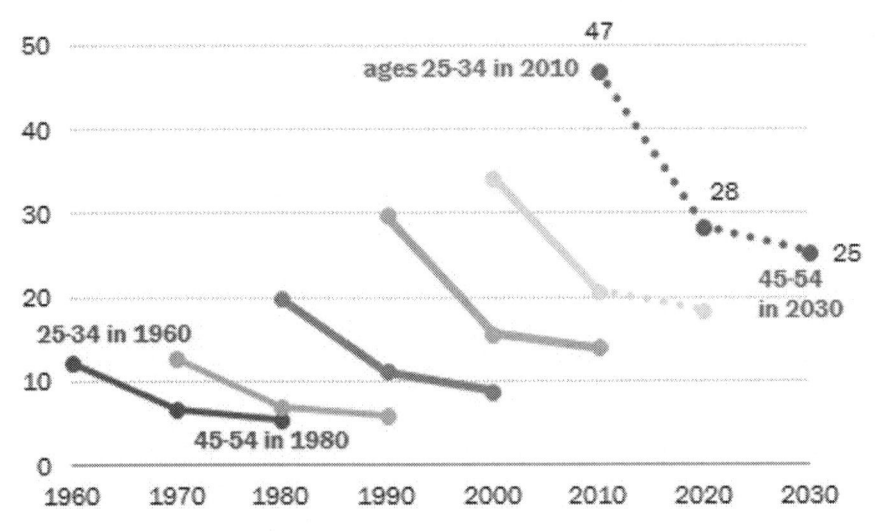

Note: The dotted lines are projected rates based on rates of the previous cohort.

Source: Pew Research Center analysis of the Decennial Census and American Community Surveys (ACS), IPUMS

PEW RESEARCH CENTER

Figure 10 More young adults are not getting married. The trend indicates the percentage of young adults that will marry is decreasing. Note, the dotted lines indicate projections based on the patterns of each cohort.

The Pew study reveals the disturbing trends.[19] Sixty years ago, only slightly more than ten percent of adults aged 25-34 were not

---

[19] Wendy Wang and Kim Parker, "Record Share of Americans Have Never Married", Pew Research Center, September 24, 2014, Source: Pew Research Center analysis of the Decennial Census and American Community Surveys (ACS), IPUMS, http://www.pewsocialtrends.org/2014/09/24/record-share-of-americans-have-never-married/, accessed June 21, 2018.

The Crisis of Identity

married. Of those not married, one-half would eventually become married by the time they reached the age of 45-54. From 1960 until today, this percentage of unmarried adults has grown to almost half of the population of the young adults. This group is projected to have one-fourth of their numbers still single when they enter mid-life, ages 45-54.

In the United States, the percentage of adults who were married in 1960 was nearly 70%. That fraction has slowly decreased to today's 54%. During the same period, the percentage of never

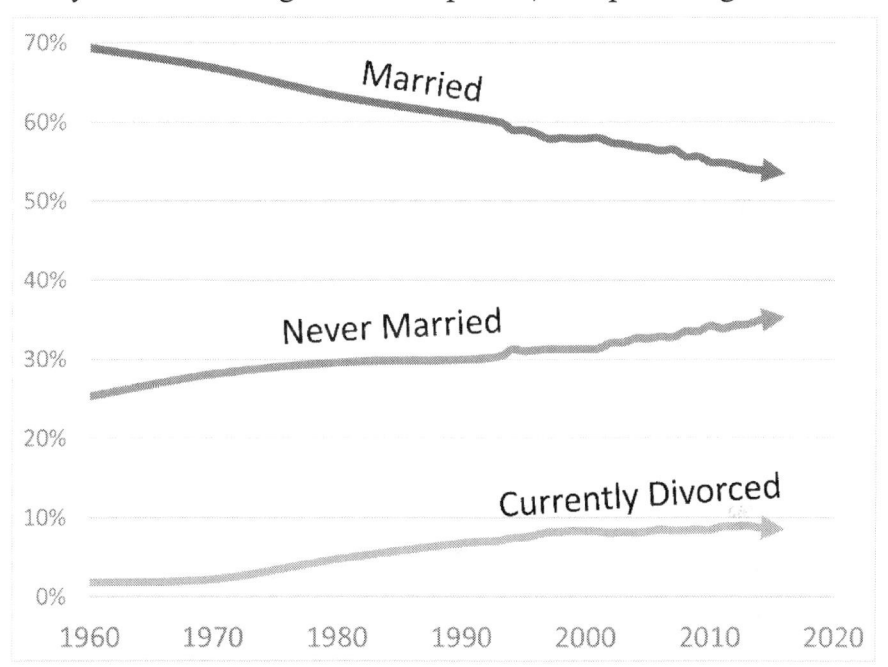

Figure 11 Comparison of adults in the United States population who were married, never married, and divorced from 1960 to present. Data from US Census and American Community Survey.

married adults has steadily increased. We should expect these trends to continue. As they do, married adults will become a minority in less than seven years (2025), and never married adults will outnumber married adults by 2050.[20] We do not have to wait until then to realize marriage no longer defines or is an important component to family relationships in our culture. The impact on our culture is devastating.

The crisis of marriage and impact on the family is fully on us. As the church has increasingly neglected the importance of marriage and families, the fruit of that neglect has come to maturity. There are churches with large percentages of their single young men and single young women who have no thought of marriage or family.

As a professor in a Christian college, a major portion of my work is formal and informal counseling. It is a significant component of teaching to help students develop and implement plans for their future. With only extremely limited exception, when asked about marriage, these students simply have no plans. They haven't given marriage even a passing thought. If they have, As a professor in a Christian college, a major portion of my work is formal and informal counseling. It is a significant component of teaching to help students develop skills and implement plans for future success. With only extremely limited exception, when asked about marriage, these students simply have no plans. They haven't given marriage a passing thought. If they have, they are hesitant to admit it. Almost as if they are ashamed to have considered marriage at an early age. Getting married and raising Godly children is not part of their view of the present or future. Only finishing college and getting a well-paying job. Their priority is self-fulfillment. Their goal is self-realization. Even those who

---

[20] Trends such as these are generally difficult to predict. However, I suspect these trends will not continue. They will change in such a way that married adult will be outnumbered by never married much sooner. In other words, the rate of change will not remain the same, but will increase.

The Crisis of Identity

proclaim a ministry or missionary profession still desire to do so based primarily in self-focused reasoning. What happened?

The answer is that we have a crisis of identity. We do not know *who we are*. Part of that crisis comes from viewing the Genesis account of creation which contains the establishment of marriage as just another story. Even though God placed the male and female together in relationship, establishing marriage before any other human institution (before church, work, club, association, government, etc.), and did so in real space and real time, in our neglect, we have chosen to make it less important. What should a young person of marriageable age do? If your answer is "go into missions work" at the expense of marriage, you need to understand this represents an unbiblical view of marriage. If your answer is "education, or work, or any other ministry" at the expense of marriage, you need to understand this represents an unbiblical view of marriage.

I know at least one person reading this has already formed the "but...". We want to define the truth based on the exception. Our imbedded beliefs do not give way easily. Before you reject the point, make sure you are reading the point. I have not said that we shouldn't prepare ourselves for marriage by getting an education, or pursuing employment, or studying God's Word, or exposing ourselves to experiences to help us grow. However, these must be in proper priority. Each of these must be placed within a Biblical order of importance. Our education, our profession, our social status are all secondary to the mission to be image bearers of God as male and female. Education, profession, employment, and ministry opportunities are only the tools to let us carry the message of Christ's love for the church, and the Church's devotion to Christ in the relationship of husband to wife, and wife to husband.

This crisis of identity means we no longer understand the created function and design and purpose of why God created male and female. We no longer understand that the image bearing design of God includes the fact that we are male and female. We rejected the knowledge of marriage and family, and how that knowledge points to God. The crisis includes the revelation God has placed in the union and relationship between husband and wife as an example of the Savior and the Church. This is what the Apostle Paul is teaching when he wrote the following:

> *Wives, submit to your own husbands, as to the Lord. For the husband is the head of the wife even as Christ is the head of the church, his body, and is himself its Savior. Now as the church submits to Christ, so also wives should submit in everything to their husbands. Husbands, love your wives, as Christ loved the church and gave himself up for her, that he might sanctify her, having cleansed her by the washing of water with the word, so that he might present the church to himself in splendor, without spot or wrinkle or any such thing, that she might be holy and without blemish. In the same way husbands should love their wives as their own bodies. He who loves his wife loves himself.*
> *Ephesians 5:22-28*

In part, because Christians have ignored the real events of Genesis regarding the creation, because they have treated the accounts of the creation of man and woman as "doesn't really matter, since it is just a story, an allegory, a metaphor", we no longer have a real event that carries the force of truth regarding family and marriage. It is a *crisis of identity*.

Identity Denied: Gender No Longer Matches Gender

Modern culture is facing a multitude of confusing changes and trends. All of them share a common source in rebellion against God. They all derive from a crisis of authority tied to the crisis of knowing who we are is the crisis of identity. This is particularly shown in our understanding of sex. There is a trunk of confusion rooted in a denial of the created purpose of male and female.

> *He answered, "Have you not read that he who created them from the beginning made them male and female, Matthew 19:4*

Christians understand from the Gospels this is a clear statement by the Creator Himself. This is both our Savior, Jesus Christ, and our Creator speaking with ultimate authority about the creation of mankind. God created us as two distinct sexes, male and female.

Jesus was answering a question from the cultural elite, the cultural leaders. They were testing Him, or attempting to trap Him by drawing out an argument they thought would cause division in His followers, and perhaps harm his reputation. It is important to note that His answer began with the rebuke, "have you not read…". He was addressing the leaders who had the responsibility for doing what was right. These leaders who should have done what all leaders must do. They must study, learn, and strive to know what was correct and then do and teach those things. For the Pharisees, the source of knowing what was correct was as clear as the phylactery on their foreheads – the Word of God.

Jesus clearly states His intention was to create two distinct sexes. One male. And, the other the complement of male, female. The male is the complement of the female, just as the female is the complement of the male. They are created as two distinctly designed forms of one kind. We are human. Male and female.

This complement-design goes far beyond the physical difference between male and female. We are designed complementally different to the very core of our gender and sex.

Our culture has become rabid in its rejection of God's designed purpose of identity in male and female. One primary pillar of that rejection is a demand to claim that male and female identity is insufficient. The vessel has turned to the vessel maker and said, "you did not get this one right." We have turned to our Creator and proclaimed the right to define our gender and sex.[21] To accomplish this proclamation of self-authority, the culture has created new types of gender, and descriptions of sex. No longer just male and female, but now a tangled, confusing, constantly changing set of gender options. Each with their own malleable implementation of self-fulfilling sexual interaction.

cisgender agender transsexual intersex cisgender male cistembender cisgender female transwoman trans person gender fluid neither nonbinary trans man gender neutrois trans masculine genderqueer transgender androgyne cissexual

This is a crisis of identity beyond full bloom and into decaying fruit. Not only is the culture claiming the right to choose a gender

---

[21] Gender is distinguished from sex in this use as: gender is a person's sexual identity, or how they self-identify. Sex is a biological state. A cisgender person has a sexual identity that matches their biology, while a transgender person has a sexual identity the is the opposite of their biology.

The Crisis of Identity

beyond the two assigned by the Creator and defined by our physically clear sex, this same culture demands the right to change gender identity as easily as changing one's clothes. The culture demands that we must see and accept a person who identifies as "gender questioning" today and "bigender" tomorrow as normal. A Christian who claims there are only two sexes is viewed as abnormal and dangerous to culture.

Being confused is not limited to the God-rejecting culture. Self-proclaiming Christians are also falling into this crisis of identity.

One family recently made the headlines by sharing their story as, "I Had 4 Boys – Until One of Them Told Me She Was Really a Girl".[22] A mother who claims to have been raised a devout conservative Christian, and active member of a local church, shares her story of her child born male exhibiting "strong female characteristics" at 18 months of age. She states that after struggling with the change, and receiving advice from a child psychiatrist, she finally realized the only thing keeping her from allowing her son to be a female was peer pressure. Not truth. Not God's Word. Not her son's biology. But, peer pressure. The dominating force that originally caused her to fear allowing her son to be female, and that changed her mind was the culture. This mother, attempting to further justify her action reaches out for a Biblical reason. She rationalizes that by studying the interactions of Jesus with the Pharisees, it became clear to her the Jesus confronted hate by asking others to view scripture "from the perspective of loving the person."

This crisis of identity, not knowing who we are, is linked to a crisis of knowing. More on that in the next chapter.

---

[22] By Kimberly Shappley, as told to Breanne Randall, *Good Housekeeping* online, Apr 13, 2017, https://www.goodhousekeeping.com/life/parenting/a43702/transgender-child-kimberly-shappley/, accessed June 27, 2018.

What do we say to this mom? Do we give in to the confusion? Will that be the "best" we can give her? Do we rationalize that what she really needs is affirmation and loving agreement with her decision? Or, is there a higher authority who has created clearly what He has created? If this mother is correct in her thinking, then somewhere along the way, God made a mistake, the Bible cannot be trusted, and who we are as male and female has no meaning based in physical reality.

What did this mother base her decision on? Was it on truth clearly, unambiguously, unapologetically taught to her by her church? She did not decide to raise her child as a girl because it was the right thing to do as clearly taught as transcendent truth. No. It was something else. She had no identity to stand on.

Our identity has become confused. We no longer know that in real space, at a real definite physical location, at a specified period of real time, God created a male and a female. He created "male and female", the identity of two separate and identifiable sexes with specific function and purpose. Who we are as created by God is based in truth that is founded in God's Word and specifically and particularly in Genesis.

Identity Affirmed or Identity Denied

Christians must be clear and precise in our identity of image bearing. We must speak to ourselves and to one-another as well as the culture in which we have been placed. The message we carry includes the clear image of God that we are. We must carry the

The Crisis of Identity

identity that God has ordained in His purposeful, specific creative act. Attempts to incorporate any type of evolutionary component confuses the incorporation of being image bearers of our Creator.

> *So God created man in his own image, in the image of God he created him; male and female he created them. Genesis 1:27*
>
> *So the LORD God caused a deep sleep to fall upon the man, and while he slept took one of his ribs and closed up its place with flesh. And the rib that the LORD God had taken from the man he made into a woman and brought her to the man. Then the man said, "This at last is bone of my bones and flesh of my flesh; she shall be called Woman, because she was taken out of Man."*
> *Genesis 2:21-23*
>
> *He answered, "Have you not read that he who created them from the beginning made them male and female, and said, 'Therefore a man shall leave his father and his mother and hold fast to his wife, and the two shall become one flesh'?*
> *Matthew 19:4-5*

Does it matter if we are young earth or old earth? The answer is a hope-giving, truth-affirming, God-worshipping "yes".

Young-earth places the specific creative act in a real definable place at a real definable time. The creation of male and female, of Adam and Eve, real people. Real persons created with purpose and tasked to carry the image of God.

We must be clear and precise in our identity as image bearers of God. In the identity God has ordained. His purposeful specific creative act. Not produced by unguided, chaotic, chance events as required by evolution.

This Biblically derived anthropology that is anchored in a real creation event confesses the all humans are:

- One Race.
- Created in God's image and bearing that image.
- Two distinct, fixed, and complementary sexes.
- Distinct and different from the rest of the creation.
- Placed in families as the foundations of culture.
- Tasked with a proper human dominion and stewardship over all of creation.

A humanistic view of anthropology, based in an evolutionary understanding of our origin, rejects the value of humans and perverts the relationships of humans to themselves and to the rest of creation. In our sinful, fallen state, combined with a belief that humans are nothing more than the end result of an evolutionary process, and only the highest form of animals, evolutionary anthropology will allow for:

- Race-based slavery.
- Abortion as birth control.
- Eugenics.
- Genocide.
- Euthanasia.
- Gender as a personal and flexible choice.
- Sexual expression based primarily in self-fulfillment.
- Family minimized or opposed.

The Crisis of Identity

Figure 7 Still photograph from the Soviet Film of the liberation of Auschwitz, taken by the film unit of the First Ukrainian Front, shot over a period of several months beginning on January 27, 1945 by Alexander Voronzow and others in his group. Child survivors of Auschwitz, wearing adult-size prisoner jackets, stand behind a barbed wire fence. Among those pictured are Tomasz Szwarz; Alicja Gruenbaum; Solomon Rozalin; Gita Sztrauss; Wiera Sadler; Marta Wiess; Boro Eksztein; Josef Rozenwaser; Rafael Szlezinger; Gabriel Nejman; Adek Apfelbaum; Hillik (later Harold) Apfelbaum; Mark Berkowitz (a twin); Pesa Balter; Rut Muszkies (later Webber); Miriam Friedman; and twins Miriam Mozes and Eva Mozes wearing knitted hats. Public Domain.

Adolph Hitler in his establishment of a New World Order, or Germany's Third Reich, desired to help evolution along by eliminating "unfavored races" and "unfavorable individuals". Sir Arthur Keith, a scientist who lived in Great Britain during the Second World War, and a proponent of "scientific racism" himself, described the relationship between Hitler's Third Reich and Darwin's evolutionary theories.

> "The German Führer, as I have consistently maintained, is an evolutionist; he has consciously sought to make the practice of Germany conform to the theory of evolution. He has failed, not

because the theory of evolution is false, but because he has made three fatal blunders in its application. The first was in forcing the pace of evolution among his own people; he raised their warlike passions to such a heat that the only relief possible was that of aggressive war. His second mistake lay in his misconception of the evolutionary value of power. All that a sane evolutionist demands of power is that it should be sufficient to guarantee the security of a nation; more than that is an evolutionary abuse of power. When Hitler set out to conquer Europe, he had entered on that course which brought about the evolutionary destruction of Genghis Khan and his Mongol hordes (see Chapter 34). His third and greatest mistake was his failure to realize that such a monopoly of power meant insecurity for Britain, Russia, and America. His three great antagonists, although they do not preach the doctrine of evolution, are very consistent exponents of its tenets."[23]

Darwin also made clear statements that devalued specific human groups, relating black-skinned humans as being more closely related to apes. He proposed that these "savage races" would be exterminated and replaced by the more "civilized races".

The great break in the organic chain between man and his nearest allies, which cannot be bridged over by any extinct or living species, has often been advanced as a grave objection to the belief that man is descended from some lower form; but this objection will not appear of much weight to those who, from general reasons, believe in the

---

[23] Sir Arthur Keith FRS, *Essays on Human Evolution*, (London: Watts & Co., 1946), 210 (cf. Evolution and Ethics, (New York: G. P. Putnam's Sons, 1947), 229.

The Crisis of Identity

general principle of evolution. Breaks often occur in all parts of the series, some being wide, sharp and defined, others less so in various degrees; as between the orang and its nearest allies— between the Tarsius and the other Lemuridae between the elephant, and in a more striking manner between the Ornithorhynchus or Echidna, and all other mammals. But these breaks depend merely on the number of related forms which have become extinct. At some future period, not very distant as measured by centuries, the civilised races of man will almost certainly exterminate, and replace, the savage races throughout the world. At the same time the anthropomorphous apes, as Professor Schaaffhausen has remarked, will no doubt be exterminated. The break between man and his nearest allies will then be wider, for it will intervene between man in a more civilised state, as we may hope, even than the Caucasian, and some ape as low as a baboon, instead of as now between the negro or Australian and the gorilla.[24]

Modern Western culture has rejected these horrifically wrong thoughts. It has done so from a basis of changing social construction. Christians can reject the separation of humanity into hierarchical races by acknowledging the real creation of a real man and a real woman on the real sixth day of creation. We are all one race of image bearers of God.

It matters what we believe about Genesis. A real creation event establishes the purpose of God. One human race. All image bearers of God. *It matters.*

---

[24] Charles Darwin, *The Descent of Man and Selection in Relation to Sex* (1871), John Murray, Albemarle Street, London, Volume I, Chapter VI: "On the Affinities and Genealogy of Man", pages 200–201.

CHAPTER 4
THE CRISIS OF UNDERSTANDING

"In his book Evangelicalism in Modern Britain: A History from the 1730s to the 1980s, *British historian David Bebbington provided a definition of evangelicalism that has become the standard boilerplate understanding for academics and journalists on both sides of the Atlantic. He described evangelicalism in terms of four distinctives:*

biblicism (a confidence that the Bible is the Word of God),

conversionism (a belief that persons must come to a saving knowledge of the Lord Jesus Christ),

crucicentrism (a belief that the cross and the resurrection are the central acts whereby God saves sinners), and

activism (evangelicals are people who hold crusades, build colleges and seminaries, go on mission trips, organize conferences, and create periodicals and publishing houses).

> *Noticeably absent from Bebbington's list, however, is the idea that evangelicals are defined by their thinking.* ***This is to our shame.****"[25]*

Christians should be careful and dedicated thinkers. Not dedicated to thinking, as in making intellectual activity our purpose or highest goal. When that occurs, knowledge and intellectual achievement have become idols. We end up making knowledge and intellect our gods. They are not. The chief purpose of mankind is to glorify God, and enjoy Him forever.[23] Therefore, we must purposefully avoid both traps of underemphasizing and overemphasizing knowledge. We must carefully bring thinking, and thinking about thinking, into a proper worship composed of everyday living.

At the heart of *knowing* is the question, "how do we know what we know?" As we answer this question, it is an epistemological event in our lives. We come face to face with knowledge theories. With our own theory of knowledge.

How do we know that God exists? How do we know that what we believe to be true is really true?

Our worship of God must acknowledge Him as the creator of all things to include knowledge, intellect, and reason. He is the source, creator, and giver of both what is knowable and what we do know. Of what can be known and what is known. He created our minds and gave us the ability to think and process knowledge and given us the ability to gain new knowledge. Amazingly, we are not limited to only discovering existing knowledge. God has

---

[25] R. Albert Mohler, Jr., Ph.D., "The Way the World Thinks: Meeting the Natural Mind in the Mirror and in the Marketplace", in *Thinking, Loving, Doing*, John Piper and David Mathis, Eds., Crossway, Wheaton, Illinois, 2011, p. 47. Emphasis added.

given us the ability to synthesize knowledge. We can add to the manner, method, and tools of how we discover and develop understanding.

God is logical and reasonable. Because He is the source of all things, both logic and reason are from Him, also. Our ability to be logical and use reason is because God has created a logical and reasonable universe, has revealed Himself in logical and reasonable ways, and created in us the ability to be logical and reasonable while being able to comprehend logic and think reasonably. All of the functions of our minds to interact with what is knowable are designed by God to be used to the fullest extent of that design for His glory. This need to fully worship God with our bodies as well as our intellect is what Paul had in mind as he encouraged the church at Rome[26], saying;

> *I appeal to you therefore, brothers, by the mercies of God, to present your bodies as a living sacrifice, holy and acceptable to God, which is your spiritual worship. Do not be conformed to this world, but be transformed by the renewal of your mind, that by testing you may discern what is the will of God, what is good and acceptable and perfect.*
> *Romans 12:1-2*

Throughout scripture, we encounter God calling for His people to use their minds. An example is found in the prophecy of Isaiah, which begins with God confronting Israel with their self-righteousness and sin. He calls them to a test. This test is an invitation to stand before judgement. God initiates this revelatory exam by reaching out to His people using a court-like trial designed to reveal to them their self-righteousness. God is using an

---

[26] Of course we mean the city in Italy, not the one in Georgia. And, specifically, the Church as incorporated in 40-90 A.D., as opposed to any modern. Though, of course, the message is just as applicable to all of the greater Church today.

intellectual tool to help them (and us) understand self-righteousness for what it is. He makes this call by noting they will be able to participate in and understand and comprehend. They are called to reason.

> *"Come now, let us reason together, says the LORD: though your sins are like scarlet, they shall be as white as snow; though they are red like crimson, they shall become like wool. Isaiah 1:18*

This is not a call to "feel" or to "experience". It is not a call to describe our place in a community, or the importance of an existential feeling. It is a call to come to a correct knowledge.

To repent of our sin we must first comprehend that it is sin, and that it is an affront to God, and that there is a way and a To repent of our sin we must first comprehend that it is sin, and that it is an affront to God, and that there is a way and a reason to turn from sin. We must know what to turn to. Comprehension of Jesus Christ as the sinless sacrifice who gave His life in the place of ours is a necessary intellectual component of salvation. We must likewise comprehend that this same Jesus, after dying on a cross, was raised from the dead after three days and remains alive to this day.

> *For I delivered to you as of first importance what I also received: that Christ died for our sins in accordance with the Scriptures, that he was buried, that he was raised on the third day in accordance with the Scriptures,*
> *1 Corinthians 15:3-4*

These all require a level of intellectual action and comprehension of the consequences of intellect and intellectual action.

Even though highly educated Christians are well known in our culture, and many churches now desire a pastor with an advanced degree, Christians are still resistant to thinking themselves. We want our pastors to be characterized by Biblical thinking, but we don't really know how to determine if he is thinking faithfully or not. Dr. Mohler, who's quote began this chapter, makes this need and distinction very clear.

> *Children do not often think about thinking. Within the developmental stage of early adolescence, there comes a sudden acknowledgment that there are other minds. "People think differently than I, or even my parents, think," a young teen will exclaim. By adolescence, we perceive ourselves thinking and begin to think about that process. Most human beings, however, never attempt to think deeply about thinking.*
>
> *By contrast, a Christian understands that he or she was made to bring God glory, to point persons to Christ, to exalt in the things of Christ, and to meditate upon God's Word. Because of the biblical imperative to be transformed by the renewing of our minds, Christians must perpetually think about thinking. Philosophers call this a "second-order discipline." Thinking is a first-order discipline, but thinking about thinking is a second-order discipline. This complex thinking is required if we are to measure and contrast faithful thinking over against unfaithful thinking.[27]*

To be the Gospel-centric activists that Christians are truly called to be, to be the Gospel-carrying and transmitting people we are commissioned to be, to be the image-bearers we are created to be,

---

[27] R. Albert Mohler, Jr., Ph.D., "The Way the World Thinks: Meeting the Natural Mind in the Mirror and in the Marketplace", in *Thinking, Loving, Doing*, John Piper and David Mathis, Eds., Crossway, Wheaton, Illinois, 2011, p. 48.

we must think rightly, and we must think about thinking rightly. And, we must be able to discern between right thinking and wrong thinking. Yet, this is often the last thing Christians think about, if at all.

Even the most worship-minded Christians who are actively working to be intellectually fully engaged have an uphill battle to properly use our minds. Why? Though God created us with a mind and intellect perfect according to His creation, these were negatively affected by the fall. That's right. Even our ability to think has been impacted when Adam sinned. Our intellect and our minds are all changed by the effects of sin. Before we can hope to think in a way that has a positive impact on ourselves and on our culture, we must be able to acknowledge the noetic effects of the fall. If we refuse to recognize the mountain before us, we have no hope of climbing. If we choose to ignore the effect of sin both entering the cosmos and still at work on our mental ability and intellect, we are hopelessly lost with no means of relief.

What's the point? What does this have to do with young-earth creation and the literal and historical events of Genesis? Before we can recognize the impact of the consequences of what we believe about Genesis and the creation, we must first account for the very effect the events described in Genesis have on our ability to think about those events. If you think on that for any length of time, it may make your head hurt.

Therefore, before we can see the consequences of a wrong view of creation on our thinking, we must account for and understand the barriers to thinking about thinking. The first barrier is perhaps the greatest. All the other effects flow from this one effect. And, all we need to do is return to the Apostle Paul and his letter to the Church at Rome.

> *For the wrath of God is revealed from heaven against all ungodliness and unrighteousness of*

The Crisis of Understanding

> *men, who by their unrighteousness suppress the truth. For what can be known about God is plain to them, because God has shown it to them. For his invisible attributes, namely, his eternal power and divine nature, have been clearly perceived, ever since the creation of the world, in the things that have been made. So they are without excuse. For although they knew God, they did not honor him as God or give thanks to him, but they became futile in their thinking, and their foolish hearts were darkened. Claiming to be wise, they became fools, and exchanged the glory of the immortal God for images resembling mortal man and birds and animals and creeping things. Romans 1:18-23*

The principal noetic effect of the fall is that humans in community and humans individually are actively suppressing the truth God has revealed. The English Standard Version uses the word "suppress" for the Greek word κατέχω (*kat-ekh'-o*), which means to hold down, restrain, hinder. Adam Clarke, a British Methodist Theologian (1760-1832) and author of a well-known commentary on the Scriptures, suggests this "should be translated, 'who through maliciousness hinder the truth.'" This is a very good understanding of the challenge we face in ourselves when we work under the burden of knowledge. It is also important when we are laboring in the fields of the Gospel with those who have not been regenerated by the Holy Spirit. People are actively suppressing the truth. John MacArthur reminds us they do this with purpose:

> *"Sinful men oppose the idea of a holy God because they innately realize that such a God would hold them accountable for the sins they love and do not*

want to relinquish."[28]

Even the regenerated person still struggles with the sin nature that is being redeemed by the work of the Holy Spirit. Christians who have worked long at submitting their minds to the changing truth of God's Word effectively applied by the Holy Spirit will still need to watch, repent, and change their thinking that may be suppressing God's truth. Only by having a mind to be on guard can we begin hoping to not be forever unable to come to the truth.

Our sinful nature works against the revelation of God. Against knowing the truth God has given. Until we can acknowledge this, we have no hope of working in the opposite direction. No hope of moving from rebellion, to the truth, to acceptance of the truth.

We've made an important first step. By realizing the in-grained active push against God's revealed truth, we can begin to account for the effects of that rebellion, and repent. Actually, remain repentant. Be constantly repenting, as in "turning away" from the active suppression of truth and turn to accepting and realizing the treasure God has given in truth. That is the first step.

A second step is to watch for the characteristics of the noetic effect of the fall and turn away from those. When we begin to think carefully about our mental rebellion and its effects, we can begin to counter the other characteristics of sin impacting our thinking. These consequences of sin can be categorized in specific areas. Knowing these can be a great help to becoming better worshippers

---

[28] John MacArthur, *The MacArthur New Testament Commentary, Romans 1-8*, Moody Press, Chicago, 1991, p. 68.

in our thinking. This list is derived from one developed by R. Albert Mohler:[29]

Ignorance

Not only are there things we do not know, there are things we do not know we do not know. We must be humble in what we do know so that we can see knowledge we are missing. One seen, we must work to fill in needed deficiencies in knowledge. We must work to learn.

Distractedness

We are easily distracted from doing the hard work of thinking. I encounter this in my own life, and see my students struggle with this daily. We have a desire to take the entertaining path instead of the fruitful path. Particularly when the path of fruit is a mental uphill climb.

Forgetfulness

Like all the others, there are dimensional aspects of this one. However, unlike the others, the intentional forgetfulness is often denied. We don't want to think we could be guilty of purposefully forgetting something. Yet, we do. We tell ourselves it is just innocent neglectfulness when we forget. Try that on your spouse for your anniversary, or child's birthday, or any other day that is about someone else. We cannot excuse our poor thinking by claiming "I forgot". No, you chose not to remember.

Prejudice

Our pre-judging of anything or anyone is almost always based in lazy thinking or sinful thinking. It is also often not acknowledged that we are being prejudiced. Because, frankly, we haven't thought about it.

---

[29] Portion of this list were taken from R. Albert Mohler, Jr., Ph.D., "The Way the World Thinks: Meeting the Natural Mind in the Mirror and in the Marketplace", in *Thinking, Loving, Doing*, John Piper and David Mathis, Eds., Crossway, Wheaton, Illinois, 2011, pp. 56-58.

Faulty perspective

Similar to prejudice, but more innocent. It is hard to constantly be on guard for thinking that begins and ends on the wrong foundation. The preeminent effect of a faulty perspective is to always find what we are looking for, instead of what we should actually have found. We choose a world-view, and then produce ideas based on that worldview. We assume ideas are correct because they are consistent with our worldview. It does not matter that the worldview may be faulty and produce wrong thinking.

Alan Bloom's *The Closing of the American Mind* warned of the change in perspective of our culture as he saw a change in college students from the 1950's to the 1960's. The 1950's student came seeking a liberal education that exposed the individual to new ideas that challenged their existing and stable philosophy. These students digested the ideas and grew in their thinking. The 1960's and beyond student came for a liberalizing of moral, ethical, and religious ideas. They were more willing to abandon existing moral thinking for the sake of liberalized sexual experience, confusing personal physical fulfillment with intellectual progress. The worldview shifted in such a way to incorporated liberal living with liberal thinking. The result was they found what they were looking for. A rationalization of any thought, any idea, as long as it was thought. That's all that mattered.

However, Does It Matter? what we believe about the age of the earth? For that scientist who has already assumed the universe is 15 billion years old, they are looking for evidence consistent with that age, and reject any evidence that points to an age that is much younger. They will find what they are looking for. Not necessarily what is really there.

Intellectual fatigue

We tire of thinking. We don't always acknowledge that thinking is hard work. It is hard. Because we do not regard thinking as hard work, we do not recognize the need to be prepared for, persevere through, or incorporate accommodations for becoming mentally "worn out". When our brains experience fatigue, just as a muscle can use up its food reserves and need to be replenished by the bodies metabolic systems, the brain, too, needs time to replenish and reset. The "pain" or "brain cramps" we feel are often interpreted as meaning "it's time to quit doing this forever." Just as important is to not push ourselves beyond what we can do effectively.

When we are thinking – like right now – we must give ourselves the grace to rest. But, we must also have the discipline to come back to the work and start where we left off.

Inconsistencies

It's embarrassing to be inconsistent in our thinking. And, it should be. However, our culture has embraced inconsistent thinking as a virtue. The post-modern and secular worldviews have combined to make the simultaneous holding of mutually exclusive ideas acceptable. For example, proclaiming a woman has the right to choose an abortion because she is a person who needs self-determination, while denying that the person growing inside her is being denied the same right to self-determination. Or, the call for totally open borders in a country actually means that country no longer exists. To be a country, there must be an identity and definition of what that nation or country is. That definition includes known borders and definable citizens. Removing that definition removes there very country people are attempting to come to.

Inability to come to the right conclusion

This is the natural outcome of suppressing the truth in unrighteousness. Genuine humility in our thinking will acknowledge that on our own, using our own abilities and intellect, we can never come to the right conclusion. Rene

Descartes failed miserably at this. Descartes declared he could not trust any external thing. Could not trust any external to him was real. The only trustworthy thing he had was his own intellect. Hence, *"dubito, ergo cogito, ergo sum"*, his most defining phrase, meaning, "I doubt, therefore I think, therefore I am."[30] Yet, his philosophy produced a major shift in Western thought leading to our current cultural worldview.

Intellectual apathy and laziness

This one is hard to recognize, and takes at least two forms. One is characterized by filling our lives with all the other things of living to the point that we have no time to grow in the knowledge of God and in the knowledge of the cosmos He has created. The Christian who is intellectually apathetic in this way knows they should study both God's Word and the creation but chooses entertainment instead.

The other is characterized by an apathetic Christian who simply isn't interested. They really could not care less to learn the science, or the math, or the literature, or the many other things that are revealing God's creation. They are too busy, in their minds, living a life focused on "heavenly things" like community and missions to be distracted away. It is possible they have relegated thinking and knowledge into a bin they label as "Secular Stuff". Not worth their time.

And, there are the Christians who simply do not want to do the work. Very likely, they struggled with math or science, or some other knowledge during their education, and out of that

---

[30] This is an expansion of Descartes actual words, and is first attributed to Antoine Léonard Thomas (1732 – 1785), a poet and literary critic (from Wikipedia contributors, "Antoine Léonard Thomas," Wikipedia, The Free Encyclopedia, https://en.wikipedia.org/w/index.php?title=Antoine_L%C3%A9onard_Thomas&oldid =844830054 (accessed July 5, 2018).

experience, "walled off" learning to keep them safe from the fear of failure.

We need to recognize and repent of all of these.

Intellectual pride

How many of us have not had to fight against this sin? It is easy to view what we think as being the best thinking. We don't want our thinking challenged. We don't want what we believe to be true to be discovered that it is wrong.

Or, after long periods of study and intellectual growth, we look at what we hold to be true and correct, and are proud that we have arrived at these truths. That pride is like all pride. It produces blindness. Because of the sinful nature in all of us, there are thoughts we are proud we have thought. This blindness causes us to stop thinking, and to believe our thinking is complete. We reach a conclusion and believe that conclusion represents the totality of what needs to be thought on a particular issue. When, in reality, we have stopped short of thinking all the way through an issue.

Intellectual fear

It may be a bit ironic that intellectual fear often comes from improper thinking about thinking.

This is not the same as the experience recounted above of the person who struggled with learning and is simply afraid of ridicule or embarrassment associated with that struggle. This is the Christian who believes knowledge is evil, and therefore chooses to be ignorant for reasons best described as pharisaical. They have been wrongly taught that knowledge always produces apostasy. Therefore, they choose to place their hope and faith in the work of ignorance. While apostasy can result from a wrong view of knowledge, it is not the knowledge that causes the falling away.

Dogmatism and closed-mindedness

There are truths we must be tenacious about. Revealed truth that is clearly identifiable as non-negotiable. Yet, the noetic effects of the fall can produce a dogmatism on issues that we have strongly linked to our feelings. We become closed-minded because we cannot think as we should. Instead of doing the hard work of careful thinking, we think to the first point of emotional comfort and then refuse to think any more. We close our minds.

This effect of the fall on our intellect has become dominant in our culture under the guise of tolerance. The "tolerant" culture has shown increasing intolerance. It is an intentional closed-mindedness that sets a boundary of acceptable thought and then actively suppresses any ideas not in full agreement with those thoughts.

Vain thinking

As a result of a sinful nature, we think about things that have no purpose, no value, no intellectual nourishment. Yet, we will allow this thinking to consume us or even control us. This is thinking, real honest mental work, that produces nothing. Or, produces ideas that are useless. These are the philosophers who produce humanistic ideas that has led the followers to despair.

Common areas include those where humans have elevated human thinking to be on par with scripture. Or, more insidiously, we claim scripture is insufficient, needing human intellectual tools to "fill in the gaps".

Miscommunication

In fear of speaking in a way that is not understood, I will not expand on this one. (I hope you appreciated the sarcasm and humor of that statement). There isn't a single person living in community who has not experienced this as both the sender and receiver of ideas that have been misunderstood, or poorly transmitted. It is a constant problem we must be aware of and take action to work against. The most important tool is to choose to be precise in our use of words. We must choose to be accurate in selecting and using words for what the mean. We need to remember that words have meaning. Meaning conveys ideas. Ideas have consequences. Therefore, we must understand that using the proper word to convey the correct meaning with full understanding that the idea being conveyed has consequences is critical to the work of preserving proper communication.

These are all challenges to good thinking. We can be aware of them and take steps to counter them. A good thinker – remember, that is what Christians must aspire too – must do. However, there are new challenges that have arisen because Christians have chosen to believe that what they believe about Genesis and the age of the earth doesn't matter.

God's Word Affirmed: What we know has an anchor in Real Space and Real Time

The stories and historical accounts in God's Word have an impact on our lives because they are true. The lessons they transmit are impactful because they come from God. But, they also have power and authority because they are real events that occurred in real space and on a real time-scale. They really happened.

Figure 8 Daniel Answers the King, by Briton Riviere (1840-1920). Note the accurate portrayal of Daniel's age by the artist.[31]

Daniel was taken into captivity in Babylon[32] where he really was thrown into a den of lions and survived by God's miraculous intervention.[33] David really did go to battle against the Philistines.[34] He really did see another man's wife bathing and spent too much time thinking and lusting after her, resulting in his fathering a child with her and arranging to have her husband killed in battle to cover his sin.[35] Joseph was actually sold into slavery,[36] taken to Egypt, and rose to become a great leader in that county and provide for his family during a time of great famine.[37]

Abraham was a real person who lived in a real city called Ur. He left that city an became the father of a new people. Noah built

---

[31] Briton Riviere, Daniel Answers the King, oil on canvas painting, 1890, Manchester Art Gallery, UK. Public Domain.
[32] Daniel 1:1-6
[33] Daniel 6, Hebrews 11:33
[34] 1 Samuel 17, 18, 23, 29; 1 Chronicles 11, 12, 14; Psalm 56
[35] 2 Samuel 11, 12; 1 Kings 1: Psalm 51
[36] Genesis 37:28;
[37] Genesis 41:41-46, 47:23-26

a real wooden craft and survived a world-wide catastrophe by God's provision of salvation. Enoch lived a real life in a real place, and was really taken up to heaven before he died. And, Adam and Eve were the first real man and woman. The first people of which there were no "pre-people".[38]

What is our claim that all of these are true? And, based on that truth claim, what support do we offer? If the stories in God's Word are not real and at best are an analogy, or only for examples, or to provide types, then we can find and extract truth from any story. All stories in our human experience will carry equal authoritative weight. We remove a reason to give the Bible greater authoritative weight respecting meaning and truth. Yes, we must confess and claim that because it is "God's Word" it is authoritative. And it is authoritative for that reason. However, if the stories have no claim from real history, if the events did not occur in real space and time, they lose their impact on any narrative pulled from them.

Let's use a story to think about this. Consider a courtroom in which a legal event is occurring. Perhaps a criminal trial. In this trial, in this courtroom, are all the key pieces we associate with a legal process: judge, jury, prosecutor, defendant, and evidence. Each truth claim used to build a "case" of either innocence or guilt must be supported by evidence. Witnesses are called and sworn to speak the truth concerning their testimony. That testimony, like all the evidence, is cross-examined to determine if it is real. The witness will be asked question to support the evidence such as, "Where were you when the events you describe occurred?" Establishing a real place supports the evidence. "When did these events occur", is used to establish the events in a real timeline. The evidence is presented as real and supported by noting its placement

---

[38] Some will prefer proto-hominids or hominids. However, please allow me to use this phrase as literary device to make the point.

in real space and real time. If the evidence cannot be established as occurring in real space and in real time, it becomes suspect, and may be thrown out or discarded.

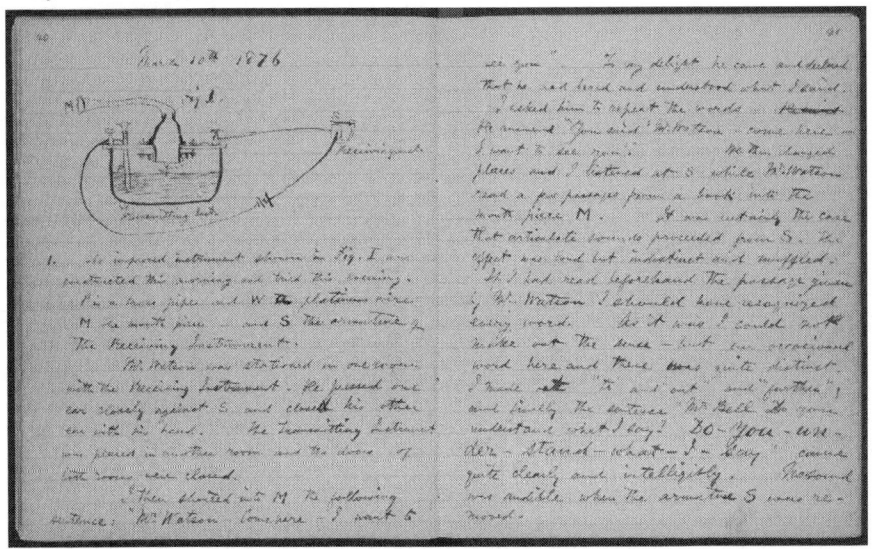

Figure 9 A portion of Alexander Graham Bell's notebook. Note the date, details journaling events that occurred. By Alexander Graham Bell - page 40-41 of Alexander Graham Bell Family Papers in the Library of Congress' Manuscript Division, Public Domain, https://commons.wikimedia.org/w/index.php?curid=7785180.

This is similar to the process of scientific work. Trustworthy science purposefully uses forensic methods from start to finish. Each step of the process from designing an experiment through gathering and analyzing the data is conducted to withstand a rigorous cross-examination. Essential to this process is recording the place and time as part of the experimental details. These details are an essential component of journaling the researcher uses to capture their experimental process. Even as they analyze and produce initial conclusions, a good researcher notes the date, time, and location. When any conclusion is offered, it must also be time-stamped. Often, a signature is included. When reporting on these events at any later time, they have the real event to "reach back too" as having actually occurred in real space and time.

The Crisis of Understanding

We celebrate birthday and wedding anniversaries because they remind us of a real event that occurred in real space at a known real time. The celebration has meaning because of the actual event being celebrated. Why would we celebrate something that did not actually occur?

Jesus lived as a physical being. He could be touched. People watched Him eat real food. Humans watched Him get tired, sleep, wakeup just like any other human does. They watched Him die. The witnessed and saw Him arisen from death. He was seen alive after being dead. We can place these events on a real timeline. We know how long ago these events occurred.

The Bible contains actual historical events. These are real events recorded as actual accounts. The narrative history in Genesis 1-11 is a real historic account describing actual events that occurred in a real space and time. These are not just good stories that teach us important life lessons.

C. S. Lewis' Space Trilogy[39] and Narnia Novels[40], or J.R.R. Tolkien's Novels of Middle Earth[41] all have great life lessons in them. But, they are not real. They are fantasy. If Lewis' and Tolkien's novels and the Bible all contain great life lessons, does it matter if any of it is real? The answer is, "Of course". First, we know that Lewis and Tolkien and others have drawn on truth from God's Word in weaving their stories of fantasy. But, they are still fantasy. Without the space-time anchor, we could use any book.

---

[39] This is a reference to the three books by Clive Staples Lewis: *Out of the Silent Planet (1938)*, *Perelandra (1943)*, and *That Hideous Strength (1945)*.

[40] A series of seven fantasy novels by C. S. Lewis: *The Lion, the Witch, and the Wardrobe* (1950), *Prince Caspian* (1951), *The Voyage of the Dawn Treader* (1952), *The Silver Chair* (1953), *The Horse and His Boy* (1954), *The Magician's Nephew* (1955), *The Last Battle* (1956).

[41] J. R. R. Tolkien published these as a prequel and three stories: *The Hobbit, The Fellowship of the Ring, The Two Towers, The Return of the King* (1954-1955).

The Bible stands apart because it is revealed truth from God. It stands apart because it contains real historical events that God designed to communicate revealed truth. The events are real. The events happened.

The resurrection of Jesus is clearly important to our faith. Paul noted that if Jesus did not rise from the dead, if our Savior had not really, actually come back to life after being really, truly dead, then our faith is in vain.

> *Now if Christ is proclaimed as raised from the dead, how can some of you say that there is no resurrection of the dead? But if there is no resurrection of the dead, then not even Christ has been raised. And if Christ has not been raised, then our preaching is in vain and your faith is in vain. We are even found to be misrepresenting God, because we testified about God that he raised Christ, whom he did not raise if it is true that the dead are not raised. For if the dead are not raised, not even Christ has been raised. And if Christ has not been raised, your faith is futile and you are still in your sins. Then those also who have fallen asleep in Christ have perished. If in Christ we have hope in this life only, we are of all people most to be pitied. But in fact Christ has been raised from the dead, the firstfruits of those who have fallen asleep. 1 Corinthians 15:12-20*

Christ's resurrection is foundational to our confidence that our sins are forgiven. That He is who He said He is - God incarnate who died to fully pay the penalty for our sin. We can hope in Him for our own coming future resurrection into eternal life with Him. We have a real confidence the penalty for our sins has been fully paid. This hope is confirmed by a real event that occurred in a known location at a knowable time - His resurrection. It really,

The Crisis of Understanding

actually occurred as recorded in the Gospels. A real event used by Paul in the letter to the Corinthian Church.

> *For I delivered to you as of first importance what I also received: that Christ died for our sins in accordance with the Scriptures, that he was buried, that he was raised on the third day in accordance with the Scriptures, 1 Corinthians 15:4*

Paul used this real event to anchor the faith of the Corinthian Christians in the Gospel. The good news of their salvation by the work of their (and our) Savior, Jesus, Christ. He not only told them the resurrection was real, but also, it was in accordance with the Scriptures. Real truth had been revealed about salvation in the Old Testament. These are the Scriptures Paul references. He may have had in mind the story of Jonah (Jonah 1:7) which Jesus Himself said was a sign of His death and resurrection.

> *But he answered them, "An evil and adulterous generation seeks for a sign, but no sign will be given to it except the sign of the prophet Jonah. For just as Jonah was three days and three nights in the belly of the great fish, so will the Son of Man be three days and three nights in the heart of the earth. The men of Nineveh will rise up at the judgment with this generation and condemn it, for they repented at the preaching of Jonah, and behold, something greater than Jonah is here. Matthew 12:39-41*

Was Jonah a real person who lived in a real location at a knowable time? Was there a real fish that swallowed him and then spit him back out after three days? Or, is that story just an allegory? A metaphor or other literary tool that didn't really happen, but was simply written by ancient near eastern shepherds who told stories to one another? No. The answer must be that Jonah was a real historical person. The account of his travels and attempts to flee

God's work through him, in which he was thrown into the sea by his shipmates and swallowed by a fish is a real, historical account. Jesus talked about that account as a real event, claiming it was a sign of the work He would do. A sign that foretold His death and resurrection. If Jonah or any part of his story isn't real, then Jesus' use of it to describe his own death and resurrection loses meaning.

The Bible contains real, historic accounts that are the anchor for revealed truth. Christians are ready to accept the New Testament accounts as real history. But, some of these same Christians will abandon Genesis as not historical. Is it because there is something in the New Testament that allows them to do so? Actually, no. It is just the opposite. In that same letter to the Church at Corinth in which Paul is encouraging Christians that the resurrection is real, he uses the names of still living (at the time of the letter) eye-witnesses to the resurrection:

> and that he appeared to Cephas, then to the twelve. Then he appeared to more than five hundred brothers at one time, most of whom are still alive, though some have fallen asleep. Then he appeared to James, then to all the apostles. Last of all, as to one untimely born, he appeared also to me. 1 Corinthians 15:5-8

And, he then takes the real historical resurrection and links it to a real historical event recorded in Genesis. Not in the later part of Genesis. In the first eleven chapters. In fact, to the events recorded in Genesis chapters 2 and 3.

> But in fact Christ has been raised from the dead, the firstfruits of those who have fallen asleep. For as by a man came death, by a man has come also the resurrection of the dead. For as in Adam all die, so also in Christ shall all be made alive.
> 1 Corinthians 15:21-22

The historical account of Adam's sin is recorded in Genesis chapter 3. A real, historical account that occurred in real physical location at a real time. If the story of Adam's sin is just allegory, just a tool of nomadic desert dwellers for describing meaning but not real events, then Paul's use of the story to note that salvation comes through this one man, Jesus Christ, just as the need for that salvation came through one man, Adam, is a lie. It is based in story that isn't real.

God's Word Denied: Unable to Think Right

Multiple challenges to truth are intersecting to produce the confusing fruit of negotiated meaning

Postmodernism, once ridiculed, has become fully integrated into the way many Christian think and speak. Secularization of our culture has produced secularized thinking within our churches and larger Christian community. Failure to give a proper value to thinking, and failure to do the hard work of thinking about thinking has yielded a church that has abandoned an essential component of our mission and given it over to the unredeemed world. The very culture that is in opposition to God has dominated the discussion of knowledge and information. Not only did the church stand passively by, in many areas, the cultural norms of postmodern thinking and meaning negotiation have become normal in the church.

It no longer matters what was meant, as long as the receiver of information can make the meaning fit into their comfort zone. As long as the receiver of information can make the communication mean what they want it to mean, instead of what it actually means, everything is "OK". Instead of words and ideas being used to express what is true, they have to be deconstructed so that their meaning can be adapted to what is relevant instead.

An encounter between John MacArthur, a conservative pastor and President of a Christian school and seminary, and Robert Schuller, former Pastor of the Crystal Cathedral, is a great illustration of the effect of these forces on thinking. Along with being a popular televangelist, motivational speaker, and author, Schuller is given credit as being the father of the seek-friendly movement in Christianity. John MacArthur described the meeting occurred while he was returning to his home from working with other Christian leaders on the topic of Biblical inerrancy. On the flight back he sat next to Schuller. They both knew of each other and were knowledgeable of their differences regarding the Word of God. During their conversations, MacArthur recalled Schuller told him, "I can say I believe the Bible and make those words mean anything I want them to mean."[42]

Robert Schuller would not be considered a conservative theologian or orthodox evangelical. Schuller's statement, though, captures the current state of confusion produced by poor thinking in evangelicalism. This confused thinking acknowledges that Genesis records that God created in six days. But, the confusion believes we can make those days say anything we want them to say. Like Robert Schuller, old-earth creation or theistic evolution claims, "I can say I believe the Bible and make those words mean anything I want them to mean." Claiming to believe the Bible while making the words mean anything we want, or need, them to mean has become far more common than many of us realize.

---

[42] Account by Jeremiah Johnson, "Inerrancy Summit, Day One", Grace To You blog, March 4, 2015, https://www.gty.org/library/blog/B150304/~, accessed July 8, 2018.

The Crisis of Understanding

Still not convinced? Consider the kerfuffle when a megachurch pastor in South Carolina preached a Christmas Eve service in which he stated that the Ten Commandments were nothing more than "ten sayings". He encouraged his congregation, followed by encouraging the watching world by posting on his blog, that we should, "look at the things we had always seen as commands (and could never keep anyway) as more like promises from God that are fulfilled in Christ."[43] Instead of:

> *"I am the LORD your God, who brought you out of the land of Egypt, out of the house of slavery. "You shall have no other gods before me. "You shall not make for yourself a carved image, or any likeness of anything that is in heaven above, or that is in the earth beneath, or that is in the water under the earth. You shall not bow down to them or serve them, for I the LORD your God am a jealous God, visiting the iniquity of the fathers on the children to the third and the fourth generation of those who hate me, but showing steadfast love to thousands of those who love me and keep my commandments. Exodus 20:1-6*

The pastor taught this should be understood as:

> *"Promise #1 – You do not have to live in constant disappointment anymore."*

Instead of:

> *"You shall not bear false witness against your neighbor. Exodus 10:16*

He taught

---

[43] Perry Noble, "The Ten Commandments: Sayings or Promises?", https://perrynoble.com/blog/the-10-commandments-sayings-or-promises, accessed July 15, 2018.

"Promise #9 – You do not have to pretend."

This is the same pastor whose church featured the song, "Highway to Hell" by the band AC/DC during an Easter Service. The church is associated with a major, main-stream protestant denomination, and in 2016 listed as the eighth fastest growing church. The pastor, after being addressed by the concerns of other church leaders proclaimed he would do it again. He defended his choices and actions by stating,

> *We all learn from the past--and I learned SUCH a valuable lesson with this one.*
> *What would I do instead?*
> *It's simple...we would have actually kicked off the service with "Hells Bells" (also by AC/DC) and allowed the "gong" to set up the song, and then done a mash up of the two songs together for a much better effect.*
> *Yes, I'm serious!*
> *We literally had a guy tell us that he decided to give his life to Christ in the service as "Highway to Hell" was being played- that he felt the Lord speak to him and say, "That's the road you are on, and you need to give your life to Me!"*
> *Just think how much better it could have been!!*[44]

Can we justify incorporating music that worships worldliness and confesses a rejection of God's truth as appropriate in a worship service simply because someone claimed to have been saved? Only if we abandon Christian thinking and the abundance of scriptural statements regarding worship, and replace Biblical thinking with pragmatism. The statement by this pastor shows how easy it is to

---

[44] Perry Noble, "Highway to Hell: I Would Do It Again!", https://perrynoble.com/blog/highway-to-hell-i-wouldnt-do-it-again, accessed July 15, 2018. Due to other events not related to this, Perry Noble was removed from being pastor at the church in July of 2016.

The Crisis of Understanding

think wrongly. It indicates the degree to which this wrong thinking in the church exists. Instead of working hard to have his thinking conformed to the Word of God, he has lazily followed the world into pragmatic thinking. This type of thinking, often described by "the ends justifies the means", makes the outcome or goal greater than Biblical truth. When this happens in a church, the church's ministry no longer emphasizes God's Word as controlling and defining their thinking and acting. Instead, the Word of God has been put under the authority of pragmatism.

I had a private conversation with a regular attender of this church regarding the Easter Service and the AC/DC song. This person would be characterized as highly educated and considers himself to be thoughtful and intelligent. He is currently pursuing a graduate degree in education. He thought the use of the song was completely acceptable and a great idea. In our discussion, I laid out concerns, including those raised by other Christian leaders, and discussed the regulative principal of worship, as well as Jesus' command to not copy the world. At that time he rejected all of these, also arguing that the ends justified the means.

If it doesn't matter what we do in our worship of God, only that people get saved, wouldn't that be clearly presented in scripture?

You may be thinking, "That was a local church. It may have been a megachurch. However, it was still not a widespread problem." Fine. Let's consider the issue of same-sex attraction. In particular the question of being same-sex attracted and being a Christian. Of several major challenges or issues that evangelicals have faced in this generation, the question of "can a same-sex attracted person also be a Christian?" is one that requires precise and accurate thinking. Unfortunately, the conversation on this issue is often characterized by feelings or emotion instead of sound thinking.

Perhaps one of the important illustrations of this crisis is in the popularity of pseudo-theology books. These books often take the form or a novel, or "personal event in the life of a person" story, and are written with an agenda to reveal or even correct theology. Even though the theology they are based on is wrong, they are still very popular among evangelicals. Whether they are the "Heaven is for Real" stories, or novels designed to teach that God loves those who suffer, self-identifying Christians are buying, reading, and promoting these books as worthwhile knowledge.

The popularity of the books and stories arises out of a feelings-first emphasis for what is true and valuable. The book about the young four-year old who claimed a visit to heaven was later made into a motion picture. Even though the young boy, now grown into adulthood, stated that none of the events in the book attributed to him actually happened.

Or, the novel made into a movie, "The Shack", described on a book-selling site with the line, "After his daughter's murder, a grieving father confronts God with desperate questions -- and finds unexpected answers -- in this riveting and deeply moving #1 NYT bestseller." The story is about a man struggling with tragic loss compounded with emotional and spiritual pain. He finds answers from a visit with God who is revealed as three persons that contradict if not refute what is revealed in God's Word. Yet, some Christians will purchase, read, and gladly accept the theology taught from this book.

This presents a real problem. Whenever we elevate our feelings to control what we believe, then what feels good or best becomes the controlling factor for what we decide is true or right or best. As Dan Phillips, a pastor, author, blogger, clearly articulates:

> *If it's all about emotion, the "discussion" is really beside the point, isn't it? Feelings are thought...well, felt... to be self-validating. After all,*

The Crisis of Understanding

you've got to follow your heart, right? And your heart is all about what you feel. Right?[45]

Human-based self-validating truth is dangerous in a way that cannot be overemphasized. Our feelings are individual. They only belong to "me". I'm the only one who can genuinely know what I am feeling. And, when I reach a conclusion by using my feelings, there is no external control to know if I am feeling correctly. It is like driving in traffic without care for using the right lanes or traffic signals. Why can't I drive anywhere as long as it feels right? Because, driving on a road surface, in a lane designated by lines painted on the road, obeying the stop signs and other controlling information will keep me safe. And alive. That's why!

The changes of belief on same-sex attraction and the Christian is a burgeoning problem within evangelicalism. Within the Western church, at least, this issue has reached every culture the church is in. As the same-sex attraction agenda has pushed and obtained full acceptance within the culture, churches have been pushed to be accepting, also. The issue of whether a person can be a Christian and be same-sex attracted became a test in many churches, as well as main-stream denominations.

This question of same-sex attraction is directly linked to whether God created male and female, and why He did so. To answer the question of "can a Christian be comfortable with an identity that is same-sex attracted", the clear Biblical answer is, "no". God designed us in the beginning to be attracted to the opposite sex. An essential part of our *imago Dei* is in being created male and female. Part of that design is that male and female fulfill the purpose of glorifying God in producing more humans. That

---

[45] Dan Phillips, https://teampyro.blogspot.com/2014/05/of-leprechauns-mermaids-and-loving.html, posted May 6, 2014, accessed July 22, 2018.

process only occurs when one sex has a desire for the opposite sex. The physical pleasure and fulfillment of sexual relations is a designed part of that process. And, by God's design, must take place in a covenant relationship we know as marriage between one man and one woman.

Beyond the *imago Dei* component of the male-female identity is the Gospel revealing design of the covenantal relationship between husband and wife. More on that later.

We face a battle for the truth every day. Christians have a commission to present the message of the Gospel to those around us. Yet, that culture no longer agrees there is absolute truth, and both the culture and much of the church no longer understands that what is true is the opposite of what is false.

The crisis of understanding is revealed by the inability to precisely speak of the difference between that which is true and that which is false. When Christians are unable to speak clearly truth and contrast it with false, it is because we have lost the ability to properly distinguish what is true and what is false. One reason this has occurred is that Christians have lost the desire to think well and think about thinking properly as related to sex, marriage, and families.

An increasing number of Christians view stories and historical events in the Bible as no more than analogy. Nothing more than a meaning buried in metaphor. The power and authority of a real event that occurred in a real place at a real time is lost. It has impacted the way they think, producing cheap ideas and contradictory conclusions.

God's Word Denied: Preemptive Surrender

All our choices and actions have consequences. Sometimes, the consequences are obvious. These obvious consequences can be small or large, have little impact or greater impact. We can decide to act in spite of the consequences because the impact is known to be insignificant, or because we are willing to accept the greater consequences. Unintended consequences, however, are entirely different. If we can truly not perceive or know the outcome of our choices, these unintended consequences are simply something we must address when they become known. The real mess are those consequences that should have been known, but we call unintended. These came about from not thinking correctly and properly about potential consequences of our decisions.

The consequences of denying the historical narrative of Genesis 1 through 11 and reclassifying it as something else is an intentional act. That intentional act either accounted for the consequences and accepted them or saw the consequences and considered them inconsequential. Whether we choose to consider the issue of the age of the earth as unimportant, and relegate it to a peripheral impact, or choose to incorporate deep time into the historical narrative and force the clear words to mean something else, the effects are directly related to the cause. The impact on our understanding, on our thinking, has produced some terrible consequences. Some of those in our children.

In 2004, Britt Beemer and Ken Ham reported on a study they had completed designed to identify the reason Christians who had been raised in Christian communities later abandoned the church when they became adults. Unknowingly, their study revealed the consequences of abandoning a Biblical intellectual position regarding the historical Genesis narrative. In a landmark study, they set out to determine why children raised in a Biblical environment and being taught in church would later leave Christianity and abandon what they had been taught. Their study incorporated children who were raised in an environment of regular consistent church attendance who no longer attended church after becoming adults. They surveyed this particular group of individuals in an attempt to determine the cause of their leaving. They wanted to know the reason for these people to leave the faith of their youth.

Their conclusion was heartbreaking. Heartbreaking because of the answer and because the consequences should have been obvious. These children who grew into adults who subsequently abandoned the faith of their childhood were given a dichotomy of an anti-intellectual church within an intellectual culture. Or, at least, a culture that offered answers that appeared to be rational and reasonable, while their church experience brushed aside questions by offering only emotional encouragement or simplistic platitude answers. The primary reason was a loss of trust. These individuals left their faith because their experience was that faith was shallow, hollow, and untrustworthy.

These are not the children of our unchurched neighbors or a pagan society. Not the children of the Mormons or Muslims or Atheists. These are the children of Evangelical Christians. Children raised in our own homes. Who sit at our tables. The ones we see every day through their young lives. The children we have been given the task to *train up in the way they should go*.[46]

---

[46] Proverbs 22:6 Train up a child in the way he should go; even when he is old he will not depart from it.

Deuteronomy 6:6-7 And these words that I command you today shall be on your heart. You shall teach them diligently to your children, and shall talk of them when you sit in your house, and when you walk by the way, and when you lie down, and when you rise.

The Crisis of Understanding

The 2004 study revealed decisions made by individuals many years previous. Decisions that had impacted generations of individuals. Their results confirmed what some suspected and corrected some previous wrong thinking.

As I finished high school and prepared for college, there were multiple times I was either directly told, or overheard conversations with a consistent theme: If you go away to college, you will fall away from church. Many of the Christian adults in my life warned me that once I went to college, they feared I would lose my faith. In their experience, they had seen other young church people go to college and leave their Christian faith. Or, if they had not directly experienced this, they had heard of someone's child who became apostate when they went to college.

But that wasn't my experience. My friends who were turning away from the church had already done so while we were still in high school. They still attended because that's what their parents told them to do.

It wasn't college that pulled children out of their faith. Ham and Beemer's research clearly showed these adults who left the church did not leave once they started college. They left, at least mentally and emotionally, when they were in middle school.[47]

The research surveyed young adults, 20-30 years old, who attended church regularly as children, but no longer did. Ninety-five percent regularly attended church in grade school. Presumably with their parents. By the time they attended high school, only 55% still attended regularly. Forty percent had already left the church. Before starting high school. It wasn't college. They were already gone.

---

[47] Middle school in this case means the public education between grade school (grades 1 – 5) and high school (grades 9-12).

Yes. The loss is heartbreaking. However, it is not the end of the arrow that should warn us. The consequences are not the problem.

Regularly Involved in a Church

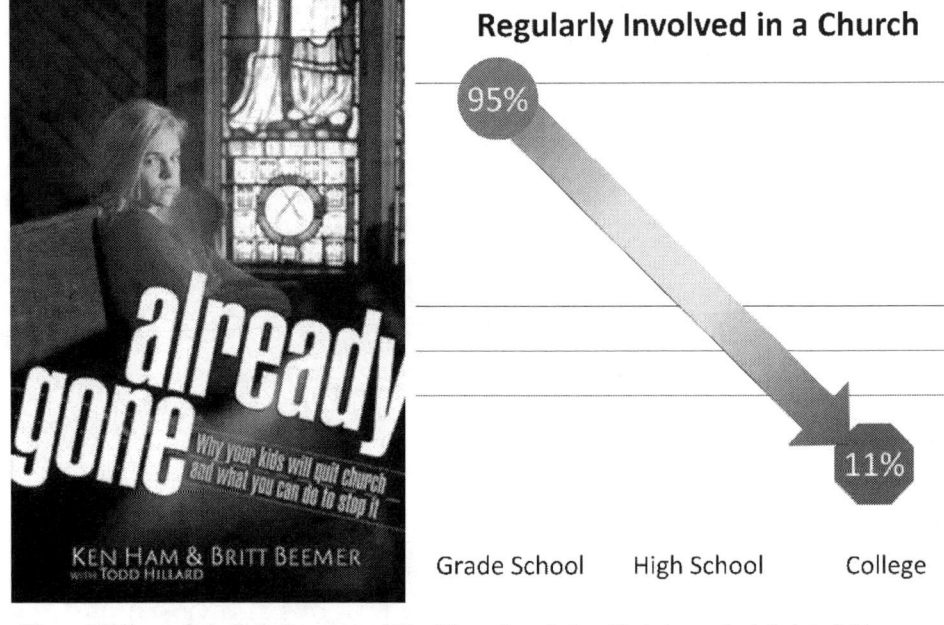

95% Grade School → 11% College (High School in between)

Figure 10 The study by Britt Beemer and Ken Ham showed that Christians who left their faith were leaving before they got to high school. They published the results and analysis of their study in the book "Already Gone".

The consequences are the effect of an action. The action is at the beginning of the trend. The cause. Something that was happening before middle and high school. Beemer and Ham reported these children were being taught by their Sunday School teachers God created the universe in six days. In school, their teachers taught the universe self-created over billions of years. The children could see these were different. The one clearly contradicted the other. They couldn't both be true. Either God created the universe as He clearly and plainly described, or evolution is correct, and the universe is self-existing and came into being billions of years ago.

When the children asked their Sunday School teachers why they should believe the Bible, they were told "Don't worry about that. Just trust Jesus." When they asked their schoolteachers why they should believe evolution, the teachers responded with evidence and reason instead of appeals to feelings. Evolution was presented as logical and reasonable, while Christian belief was presented as feelings and suspension of reason.

This study points to a clear conclusion: Believing that Genesis 1-11 is not real history of real events produces an unhealthy Christian.

In addressing the greatest questions, the ultimate questions such as, "Who Are We?", "Why are we Here?", "What is our purpose?", the answers and the process we use to find are essential to the foundation of knowledge and knowing. The only valuable answers require certainty and confidence in what is true and real from start to finish. If Genesis isn't real in the historical sense, if it isn't real in the sense of having a place in time and space, then it's only another story. It may be a better story in our minds or in our opinion, but it doesn't have the strength and force of being founded in real space and real time.

Evolution and old-earth make the claim of being real in space and time. Real events that form a foundation for the rest of thinking. This claim gives the resulting conclusions of a self-directed existence and confidence in human ideas power over revealed knowledge in other areas of thinking and understanding. This power is wrongly believed to be real. It isn't.

Young-earth creation is a confession based in real events that occur at identifiable locations in a time scale that is measurable. It confesses the revealed as rational and superior to human conclusions.

Chapter 5
The Crisis of the Gospel

Does it matter? Yes. What we believe about Genesis impacts what we know about sin and salvation. It affects the Gospel.

Allow me to offer this as a statement of *not what is to come, but what is*. Thinking and speaking clearly about sin and salvation is rooted in what you believe about the historical reality of the Genesis account.

God's Word Affirmed: Sin is Real, Knowable, Definable

Real sin became part of the real creation at a real, definable time in a real, locatable place due to the action of a real, namable person. Unless, of course, the Genesis record of these events isn't really a record at all. If it is only a narrative give to and recorded by a tribal group in what we call the Near East, and can only be properly understood as transmitting form and function, what does that do to the following statement?

> *For as by a man came death, by a man has come also the resurrection of the dead. For as in Adam all die, so also in Christ shall all be made alive.*
> *1 Corinthians 15:21-22*

Genesis contains the real account of how sin entered the world. The Crisis of Authority began when Adam and Eve challenged God's definition of right and wrong.

> *So when the woman saw that the tree was good for food, and that it was a delight to the eyes, and that the tree was to be desired to make one wise, she took of its fruit and ate, and she also gave some to her husband who was with her, and he ate. Genesis 3:6*

That real event was linked by the writer of the Letter to the Romans to another real event. The sin event which brought death to all of mankind is linked in the New Testament to the real sacrifice of Jesus Christ.

> *Therefore, just as sin came into the world through one man, and death through sin, and so death spread to all men because all sinned— for sin indeed was in the world before the law was given, but sin is not counted where there is no law. Yet death reigned from Adam to Moses, even over those whose sinning was not like the transgression of Adam, who was a type of the one who was to come. But the free gift is not like the trespass. For if many died through one man's trespass, much more have the grace of God and the free gift by the grace of that one man Jesus Christ abounded for many. And the free gift is not like the result of that one man's sin. For the judgment following one trespass brought condemnation, but the free gift following many trespasses brought justification. For if, because of one man's trespass, death reigned through that one man, much more*

> *will those who receive the abundance of grace and the free gift of righteousness reign in life through the one man Jesus Christ. Romans 5:12-17*

The denial of value and purpose of a Genesis account of creation as real history has produced confusion in the Gospel. The Gospel is in crisis.

The Gospel does not begin with repentance. There is something that comes before. It does not begin with understanding our sin or that we are sinners. There is something that comes before. The Gospel begins with understanding God. God who has authority because He is the personal creator of the universe and the human race.

Not just some esoteric concept of a "higher power". This does not mean being able to comprehend that there must be a designer based on the preponderance of design. It does mean knowing that God, as He has revealed Himself to His creation, is Whom He says He is. Only at that point does the enormity of our sin take shape. Preceding from knowing the holiness of God comes a proper understanding of the wretchedness of our sin. God who did what He did with the purpose He had is foundational to the Gospel. To all aspects of the Gospel: regeneration, sanctification, glorification.

Genesis of the Gospel Crisis

The first presentation of the Gospel is seen in Genesis 3. In the very moment the first man and woman challenged the authority of God, unleashing the curse of sin on all of creation, God applied the judgement He had warned them of while also telling them that redemption from their sin had already begun.

God confronted and revealed the sinners with their sin. Before any hope of redemption, before any promise of salvation, Adam and Eve where first shown their need. They had sinned and had become sinners in their sin. According to the historical narrative, they knew what had happened. They had surrendered to the temptation and entered fully into open rebellion against God by willfully doing what He had clearly told them not to do. Having done so, they acted like guilty children.

> *So when the woman saw that the tree was good for food, and that it was a delight to the eyes, and that the tree was to be desired to make one wise, she took of its fruit and ate, and she also gave some to her husband who was with her, and he ate. Then the eyes of both were opened, and they knew that they were naked. And they sewed fig leaves together and made themselves loincloths. And they heard the sound of the LORD God walking in the garden in the cool of the day, and the man and his wife hid themselves from the presence of the LORD God among the trees of the garden. But the LORD God called to the man and said to him, "Where are you?" And he said, "I heard the sound of you in the garden, and I was afraid, because I was naked, and I hid myself."*
> *Genesis 3:6-10*

Adam and Eve had talked with God before they sinned. They had walked with Him in the garden God specially created for them. Before they rebelled, there was no guilt, no shame, no fear. But now, Adam responds to God's question that he is afraid when before he wasn't. Adam says he is ashamed and needed to hide himself when before there was no need to hide. Adam and Eve are both aware things have changed. They already knew they had rebelled against God, and that rebellion was wrong.

God knew what they had done. And He knew they understood things had changed, and knew they understood they had disobeyed Him. Yet, God still confronts them with their sin. Clearly, purposefully, gently beginning with the fact of their sin. They were in sin.

We must then as, "Why did God ask the question?" God asked the man, Adam, to confess what had happened.

> *He said, "Who told you that you were naked?*
> *Have you eaten of the tree of which I commanded*
> *you not to eat?" Genesis 3:11*

God knew what Adam and Eve had done. The question's purpose what not to add to God's knowledge. It was for the sake of Adam and Eve. God already knew the answer. He knew they had rebelled. He wasn't asking because He wasn't there when it happened and needed to know the details. The first thing Adam and Eve needed was full realization that their actions were fully, really, totally, sin. Not some shade of "not good". Not a "stain". Not a minor stumble in their walk. Full open rebellion. Also known as *sin*.

The question confirmed in their minds what they already knew, "Yes, we did what you commanded us not to do. We rebelled against you."

"What is this that you have done?"

Genesis 3:13, How the Gospel Begins

The Gospel revealed in this historical account establishes the necessary components and order of presentation of the Gospel. You cannot begin with redemption. The message of salvation begins with understanding the need for salvation. Before we know a Savior is needed, we must fully realize that a sinner is present. We must be able to acknowledge and own our guilt. Realize it is a personal guilt. Confess the ownership of the sin.

Modern Christians have pushed this necessary order aside. We want to ease the pain of understanding in the sinner's mind of their need for salvation by softening their sin. When this happens, we do damage to the Gospel. When we attempt to make some sins "not so bad", or maybe, "not their fault", we make it harder, if impossible, for the person to then make the necessary turn away from their sin. Telling a homosexual their desire for sex with a person of the same gender is alright because they were made to have that same-sex desire throws a barrier into their understanding of their sin.

God does not leave Adam and Eve in hopelessness due to their sin. The purpose of their punishment here is twofold. It is both a just response to their rebellion and a tool designed to bring them to repentance and restoration. God renders the judgement and includes a promise.

> *I will put enmity between you and the woman, and between your offspring and her offspring; he shall bruise your head, and you shall bruise his heel."*
> *Genesis 3:15*

Theologians identify this passage as the first place in Scripture the Gospel is portrayed. It is referred to as the *protoevangelium* from two Greek words, *protos* meaning "first" and *evangelion* meaning "good news". And, it is a real event that occurred in real space at a real time.

Unless Genesis chapters one through eleven are just allegory.

Here is the Crisis of the Gospel: If the events described in Genesis 3:15 are only a story designed to tell a greater truth, then the greatest truth contained in this statement is made common. If this is just a story, then we can use any story to establish our own version of salvific truth. However, if this is a real event that occurred in real space and real time, then it exists apart from our ability to create a narrative designed to support the truth. A

The Crisis of Understanding

narrative we would design in such a manner as to give mankind priority. Or, to give something other than what God intends. This account has priority because it is real. It actually occurred. God actually spoke those real words to a real person about a real event.

The Gospel begins with an understanding of our need. No sin equals no sinner. No sinner equals no need for a Savior. There was a real first sin. It wasn't the last.

I, too, really sinned my own sin. I must repent for my sin. Without that understanding, there is no need for the rest of the Gospel message. Yet, once we have fully owned our guilt, then we can own the need for payment for the sin.

In the *First Gospel* or *protoevangelium* statement of Genesis, God continued to reveal the salvation He had determined would be made available to sinners. The next real event in that revelation is described in Genesis 3:21.

> *And the LORD God made for Adam and for his wife garments of skins and clothed them. Genesis 3:21*

God could have made the clothing from anything: any plant or natural fiber, event "vegan leather"[48]. He chose to use the skin of an animal. Although it is not specifically stated, we can understand this meant that God took an innocent animal and killed it to provide this covering. Right there in front of the sinners. In the face of Adam and Eve, God shed the blood of an innocent animal and made clothing to cover the two sinners. He provided for their need. That provision came with the shedding of blood. An animal lost its life so they could have life. An innocent life was ended to provide a covering for them. To provide a covering for their sin,

---

[48] Vegan leather was a marketing tool designed to represent material that looked like leather, but was not made from animal skin or hide. Most artificial leathers are made from polyvinylchloride (PVC), including and early form of faux leather, Naugahyde.

God shed the blood of a living creature.[49] It was instruction for the first two humans with regard to the Gospel message. This very first shedding of blood in sacrifice to supply the needs of the sinners was the first instruction for all proceeding sacrifices for sin to come.

Pause and fully consider what this means. Real blood shed means something extremely significant. A fictional story about fictional blood cannot carry the same authoritative weight of truth. Real blood shed creates a different impact on our understanding. Fictional blood in a fictional story with only a purpose to convey an idea that could be conveyed by any other literary device does not have the same impact.

Why did this have to be a real event? God could have used a literary device in any way He desired to do so. Why a real event? The answer lies in the depth of understanding of the holiness of God and the extreme insult of sin. If the sin wasn't real, but only a storyline, and the shedding of blood not real, only more of the same story line, everything is up for negotiation.

If this account is only a story, narrative, fiction, then it doesn't matter about the specifics. Any similar act could be used to provide the payment for sin.

---

[49] Adam Clarke, in his commentary on this passage, notes, "It is very likely that the skins out of which their clothing was made were taken off animals whose blood had been poured out as a sin-offering to God; for as we find Cain and Abel offering sacrifices to God, we may fairly presume that God had given them instructions on this head; nor is it likely that the notion of a sacrifice could have ever occurred to the mind of man without an express revelation from God. Hence we may safely infer, 1. That as Adam and Eve needed this clothing as soon as they fell, and death had not as yet made any ravages in the animal world, it is most likely that the skins were taken off victims offered under the direction of God himself, and in faith of Him who, in the fullness of time, was to make an atonement by his death. And it seems reasonable also that this matter should be brought about in such a way that Satan and death should have no triumph, when the very first death that took place in the world was an emblem and type of that death which should conquer Satan, destroy his empire, reconcile God to man, convert man to God, sanctify human nature, and prepare it for heaven." *Adam Clarke's Commentary on the Bible,* Adam Clarke, LL.D., F.S.A., (1715-1832), Published in 1810-1826; public domain.

The Crisis of Understanding

The story of Cain, of course, provides the necessary refutation of such thinking.

> *In the course of time Cain brought to the LORD an offering of the fruit of the ground, and Abel also brought of the firstborn of his flock and of their fat portions. And the LORD had regard for Abel and his offering, but for Cain and his offering he had no regard. So Cain was very angry, and his face fell. The LORD said to Cain, "Why are you angry, and why has your face fallen? If you do well, will you not be accepted? And if you do not do well, sin is crouching at the door. Its desire is contrary to you, but you must rule over it." Genesis 4:3-7*

Cain's offering was rejected because it was offered based on Cain's desire to control the truth instead of following God's revealed truth.

The events in Genesis 3 are real historic events that provide the authority for the truth of God's holiness, the insult of sin, and the need for a proper redemption. All contained in this real historical story.

It matters to the Gospel what you believe about Genesis.

The Gospel Demands Clarity Regarding The Need

In recent years, the Gospel crisis caused by wrong thinking about the first eleven chapters of Genesis have begun to dominate the evangelical landscape. We are going to use only a few examples to help illustrate the insidious nature of what has happened to the Gospel.

In 1889 Adoniram Judson Gordon[50] founded Boston Missionary Training Institute in Clarendon Street Baptist Church of Boston, Massachusetts. His goal was to create a school that could train Christian missionaries for work in Central Africa. The Institute flourished and grew. Over the years, this school became Gordon College. Gordon College today is a non-denominational Christian college teaching approximately 2,000 students annually.[51]

In July 2014, President Barak Obama signed Executive Order 13672 that added sexual orientation and gender identity to the list of categories protected from discrimination by federal contractors. This action effectively made sexual orientation and gender identity an identical protected class as national origin and gender. Christian colleges that either accepted federal money directly or indirectly through student funds were potentially affected by this executive action.[52] Some religious leaders responded by preemptively asking for exemption based on their religious convictions. One letter was signed by 160 individuals ranging from Presidents of Christian colleges and universities to leaders of Christian agencies. A second letter urging the inclusion of religious exemption be added to the order was sent by several religious leaders ranging from the Chief Executive Officer of Catholic

---

[50] A. J. Gordon (1836 – 1895) was named for the Particular Baptist Missionary, Adoniram Judson, who is known for his work in Southeast Asia. A. J. Gordon was a Baptist pastor, and founder of Gordon College and Gordon-Conwell Theological Seminary.

[51] Gordon College website, "History of Gordon", https://www.gordon.edu/history, accessed June 24, 2019.

[52] "Implementation of Executive Order 13672 Prohibiting Discrimination Based on Sexual Orientation and Gender Identity by Contractors and Subcontractors", https://www.federalregister.gov/documents/2014/12/09/2014-28902/implementation-of-executive-order-13672-prohibiting-discrimination-based-on-sexual-orientation-and, accessed June 21, 2019.

Charities USA to pastors and one college president: D. Michael Lindsay of Gordon College.[53]

Gordon College and President Lindsay joined the ranks of those attacked for not conforming to the social agenda of normalizing gender fluidity and same-sex attraction. The response from surrounding communities was swift as they turned against the school, with some municipalities prohibiting students from colleges associated with the request for exemption from doing work in their local public schools. The Peabody Essex Museum ended its academic relationship with the Gordon college due to the letter signed by President Lindsay.[54]

Following the initial backlash, President Lindsay issued a letter to the Gordon College community in an attempt to clarify and mitigate concerns. In summary of his intentions, the President of this Christian college that had such deep roots in spreading the Gospel ignores the essential issues of the Gospel.

> *"In general practice, Gordon tries to stay out of politically charged issues, and I sincerely regret that the intent of this [the original letter to President Obama] letter has been misconstrued, and that Gordon has been put into the spotlight in this way. My sole intention in signing this letter was to affirm the College's support of the underlying issue of religious liberty, including the right of faith-based institutions to set and adhere to standards which derive from our shared framework of faith, and which we all have chosen*

---

[53] "Religious Exemption Letter to President Obama", https://www.scribd.com/document/232327567/Religious-Exemption-Letter-to-President-Obama, accessed June 21, 2019.

[54] Neil H. Dempsey, "Peabody Essex Museum severs ties with Gordon College", The Salem News, July 25, 2014.

> *to embrace as members of the Gordon community."[55]*

The challenge brought by normalization of homosexual activity has increased in recent years. Gordon College faced two other challenges within their faculty over this issue. One from a philosophy professor who filed a lawsuit claiming she was disciplined for criticizing President Lindsay's participation in the letter 2014 letter.[56]

The Gospel should be a reason to engage on politically charged issues. Not for the sake of politics or political advancement. For the sake of the Gospel. Religious liberty is important. But not important on its own, as the statement in the letter indicates. We shouldn't take a stand on religious liberty because we support the need for "faith-based institutions to set and adhere to standards which derive from a shared framework of faith."

The Gospel identifies the needs of fallen humans: we are sinners. The Gospel clearly speaks of what sin is: open rebellion against God. The Gospel plainly identifies the remedy for our hostility toward God: His loving, perfect, complete gift of the full payment of the penalty for our offense to Him. If we lose that ability to articulate any one of these, we lose the Gospel.

Christians must agree with what Scripture says regarding God's design for the world. He created a real man and a real woman on a real day in real time. This man, a male, and this woman, a female, were told by God to act together in the manner God had

---

[55] Gordon College Website, "Questions Regarding the Letter to President Obama", https://www.gordon.edu/article.cfm?iArticleID=1625&iReferrerPageID=5, accessed June 21, 2019.

[56] Lauren Swayne Barthold, Letter to the Editor, The Salem News, July 11, 2014, http://www.salemnews.com/opinion/letter-hiring-practices-cause-pain-in-gordon-community/article_a890ec4f-d8d6-5c68-a001-5a8cd89ad042.html, and Barthold v. Gordon College, Essex Superior Court

designed to produce children.[57] Anything that denies or rebels against that created design is sin. If we cannot tell a sinner they are guilty of sin, it will make their understanding of the Gospel harder if not impossible.

The purpose of God's creation as male and female was for the two genders, both male and female, to be His image bearers.[58] We are showing the image of God in our maleness and in our femaleness. Denial of the existence of created, designed gender is a rebellion against God. Denial of a designed gender that is produce when we are conceived and born is rebellion against God. If we cannot articulate that Truth clearly to the unregenerate world, how are they to know they are guilty of sin and need a Savior? How will they know what to repent from?

Genesis is literal history. It establishes in real space and time the creation of a real man and woman by a real God Who has the authority to do what He did. That authority and design is anchored in the real events that occurred in real space and time. If Genesis is "just another narrative", we have no authority to speak plainly about sin related to gender and sex.

The Gospel is impacted by what we believe concerning Genesis.

What are We Repenting From?

Claims regarding sex in our culture are religious claims. Or, any claim purporting to be a truth statement with respect to sexual expression or practice is a religious claim. I know that seems

---

[57] "And God blessed them. And God said to them, 'Be fruitful and multiply and fill the earth and subdue it, and have dominion over the fish of the sea and over the birds of the heavens and over every living thing that moves on the earth.'" Genesis 1:28 ESV

[58] "So God created man in his own image, in the image of God he created him; male and female he created them." Genesis 1:27 ESV

difficult to comprehend. It is particularly difficult for Christians, it seems. Many Christians have agreed on some level that sexuality is a separate personal issue that can be expressed by an individual with no other impact. In their thinking, the individual expression of sexuality occurs in isolation. It's private. Whether the truth claim is a statement of gender that doesn't match their biology, or a truth claim of preferred physical expression, the only meaning of that statement is to the individual and to whomever is participating in that expression with the individual.

They are wrong.

When a Christian affirms this view, they can only do so by causing damage to the Gospel and to the message of the Gospel. Along the entire spectrum of affirmation of sexual expression that is not in alignment with God's design of a single male and a single female in a covenant relationship, the Gospel is harmed. Whether the Christian is simply silent, choosing to not speak out and not state what is clearly revealed in God's Word, or actively affirms the expression using some type of scripture-twisting thinking, it is still affirmation. It still damages the Gospel. And, that wrong thinking is often linked to a belief the Genesis account of God creating a male and female and placing His image in that creation of specific and fixed genders is nothing more than a story written by uninformed near-eastern nomadic simpletons.

Many modern evangelicals are desperately trying to keep their heads above water in the flood of LGBTQ+[59] In 2018, the first of two Revoice Conferences was held for the purpose of affirming individuals who identified as homosexual and as a Christian. The confusion of man-centered sexual identity in opposition to the

---

[59] At the time of authoring this book, this is the common way of addressing the range of sexual identities in western culture. LGBTQ+ is an acronym for lesbian, gay, bisexual, transgender, queer, and all others (plus).

clear Biblical design is seen in the Revoice Conference vision statement:

> *Revoice exists because we want to see gay, lesbian, bisexual, and other same-sex-attracted people who adhere to historic, Christian teaching about marriage and sexual expression flourish in their local faith communities. We envision a unified Church where these individuals can be transparent with their faith communities about their orientation and experience; where local churches utilize and celebrate the unique opportunities that lifelong celibate people have to serve others; where Christian leaders boast about the faith of people who are living a sacrificial obedience for the sake of the Kingdom; and where all people—regardless of their orientation or experience—are welcomed into the lives of families so that all can experience the joys, benefits, and responsibilities of kinship.*[60]

Because we are concerned about a proper view of the Gospel and the need for a proper understanding of the Gospel, we must ask, "How is it possible to 'adhere to historic, Christian teaching about marriage and sexual expression' and identify as 'gay, lesbian, bisexual, ...'?" The historic, Christian teaching affirms the Biblical statements of God's revelation of Himself is contained in the male and female. Not male alone. Not female alone and separate. But male and female. Two sexes designed for intercourse (in the larger and comprehensive meaning). Two sexes designed for life in a loving, covenantal relationship in which procreation is a major part. Two sexes designed to bear the image of God only in living together in proper covenantal submission to one another without

---

[60] Revoice website, https://revoice.us/about/our-mission-and-vision/, accessed July 4, 2019.

compromise. Striving to out-love one another as Christ loves the Church. Repenting of their selfishness and conforming to the image God determines them to be.

God's Word Affirmed: Identification of and Repentance from Sin

In Genesis, God tells us we are created male and female. That creation is purposeful revelation of God Himself. We, as male and female, are the image bearers of God. Sexual thinking and acting are all under the authoritative design and command of the Creator.

God's Word Denied: a Crisis of the Gospel

When we remove any portion of the created image of God in mankind and attempt to throw down the authority God has to establish proper, designed, fixed gender with proper, defined, and fixed sexual relationships, we confuse the need for salvation itself. If there is no sin in choosing to be whatever gender a human feels they should be, and if there is no sin and identification of sexuality as something other than God's design, there is no need for repentance.

Any form of evolution, including theistic evolution and progressive evolution, denies the literal account of God creating real individuals with real purpose. These origin stories deny the creation of specific sexes with a specific purpose and specific roles. These all deny the entrance of sin in the world that occurred as written. In a real event by the actions of the first only two humans. In effect, when the Genesis account is denied as real, there is no need for judgement because there was no real event that brought sin into the creation. No need for obedience and repentance from disobedience. According to any of these views of creation, humans can determine what is evil and what is good. There is no fixed truth.

When the Genesis account is characterized as just another story, humanity is emphasized and God is deemphasized. This demolishes the Gospel. It matters what you believe about Genesis.

CHAPTER 6

WHAT MUST WE DO?

The answer is quite simple. It doesn't require formation of large organizations. You are not going to find a strategy for organizing a movement laid out in detail. Not even a list of things to do. The answer is something you must do. It is personal and simple.

It distills down to this: *change your mind about the sufficiency of God's Word.* Confess what God clearly reveals to be historical events as historical events that have authority in our understanding. Boldly live out the truth that the events of Genesis are real historical events and the truth they convey is not negotiable or changeable to fit our modern narrative. The real events in Genesis contain truth that is the foundation for the meta-narrative of God's written revelation to us: The Bible. Those truths are necessary in today's culture. They are relevant.

Sometimes, it is hard to know how to do this. To think on how to change your mind requires you first know where you current thinking is. Once you know where you are, then you can know how to get to where you need to be. The preceding chapters where designed to show you where you are and where you need to be. However, the underlying truth remains that all of Scripture has the foundation of revealing God and His redeeming work through the Son, and does so with real, historic events that occurred in our space-time.

The crises brought about by ignoring the anchor of Truth that a literal Genesis provides has resulted in a state for the modern church that is unanchored. A church that is blown about by the wind. Very much like a church described in the Revelation.

> *"And to the angel of the church in Laodicea write: 'The words of the Amen, the faithful and true witness, the beginning of God's creation. "'I know your works: you are neither cold nor hot. Would that you were either cold or hot! So, because you are lukewarm, and neither hot nor cold, I will spit you out of my mouth. For you say, I am rich, I have prospered, and I need nothing, not realizing that you are wretched, pitiable, poor, blind, and naked. I counsel you to buy from me gold refined by fire, so that you may be rich, and white garments so that you may clothe yourself and the shame of your nakedness may not be seen, and salve to anoint your eyes, so that you may see. Those whom I love, I reprove and discipline, so be zealous and repent. Behold, I stand at the door and knock. If anyone hears my voice and opens the door, I will come in to him and eat with him, and he with me. The one who conquers, I will grant him to sit with me on my throne, as I also conquered and sat down with my Father on his throne. He who has an ear, let him hear what the Spirit says to the churches.'" Revelation 3:14-22*

When tolerance commands greater authority than Truth the result is that culture can no longer clearly identify what is truthful. Truth has been demoted to the god of cultural demands. There is no longer a trustworthy way to determine what is true and what is false. We are no longer able to know what is right and what is wrong.

Figure 11 The ruins of Laodicea. By Rjdeadly , CC BY-SA 3.0, https://commons.wikimedia.org/w/index.php?curid=19781425. Recolored.

In my lifetime, I have watched the Evangelical Church face multiple challenges brought on by a demand from the culture to redefine truth using a claim of tolerance. In the 1960's the demand was for sexual freedom. The culture cried out that sexual activity was natural and beautiful. Culture correctly identified that sexual activity can be natural and beautiful. But incorrect in their application. Sexual activity is natural and beautiful when it is placed in the covenantal union of one man and woman for life, as ordained by God. Yet, the culture twisted this by questioning why anyone would want to oppress and deny what was clearly a natural and harmless activity. Culture argued that as long as sexual activity was between consenting adults, there should be no problem. Those who took a Biblical stand were attacked by labeling them as backward, controlling, uninformed, not keeping up with the times, and just plain oppressive. Even unloving and uncaring. To oppose their view of freeing sexual activity between two people who loved each other was labeled intolerant.

Many churches gave way before this onslaught and stopped speaking the clear teaching of God's Word. They softened or stopped teaching that God had the right to establish the proper place of sexual relationships as between one man and one woman in a life-long covenant relationship. Instead of remaining bold, they became weak in emphasizing God's purpose for sexual relationships. God did not design sex simply to be pleasurable. It wasn't just for an expression of love. And, what the culture was describing as love wasn't real love. Sex has a greater purpose that God had designed.

Alongside the demand to tolerate sexual activity outside of any covenant of marriage came the demand for promotion of one gender above the other. The culture claimed men had oppressed women, and Christianity was the lead cause of this oppression.

The truth that sinful humans will oppress other humans is seen throughout God's Word. And, God's Word has clearly spoken against sinful oppression. In fact, in Judeo-Christian worldviews, woman have equal status in all areas with men. Different roles and responsibilities are defined for each sex. These roles are clearly laid out by God's clear revelation. However, the culture identified the source of the distinctives in gender as coming only from men. They reason that since we evolved from lower life forms, there can be no real need for specific roles for men or women. Particularly not anymore in our higher evolutionary state. Therefore, any roles that exist must be from oppressive males. And males had the control of religion. And in this controlling fashion, they repressed and oppressed women. Feminism responded by asking questions that should have been answered. Questions such as, "Why should women be the principal nurtures and primary caretakers of the children? Men should take an equal role!" "There should be as

The Crisis of Understanding

many women in the factories and managing the money firms as men!", was the cry from the culture. And the church gave way before them. Often, simply by remaining silent.

The church then faced the challenge of an attack on marriage. Divorce in the culture moved away from a breach of trust to the no-fault divorce. "Why can't we just leave our marriage when we fall out of love?", was the question. "Don't you Christians want people to be happy?" So, the culture was dominated by divorce and remarriage, and then no marriage at all. It wasn't long until many evangelical churches followed. The churches divorce rates and low marriage rates began to mirror the culture. Today, even evangelical churches that would consider themselves to be conservative are filled with young adults who have no plans on getting married. No plans on building families.

The last two churches our family has left was over the issue of homosexuality. Although these churches considered themselves to be doctrinally sound, even doctrinally conservative, the leadership in both embraced the idea that a person could identify as fully gay, though celibate, and simultaneously be accepted in full good fellowship with all other Christians.

Each one of these changes share a cause. Each one of the adjustments to culture the church made sprang from the same place. A loss of belief in the sufficiency of Scripture. Although if asked, the pastors and leaders in each of these churches, and particularly the last two, would say they believed Scripture was sufficient. Yet, their actions would clearly send a different message.

Doubting scripture sounds like an easy symptom to spot. It isn't. Notice the different word used to open this paragraph? We generally associate doubt with something clear and bold. Like a heretic. Or, at least, an apostate. They are easy to spot because if you ask, they will confess their doubt in God, God's Word, or some specific point of orthodoxy. However, when we speak of the sufficiency of scripture, there is already an assumption of agreement that all of Scripture is true and applicable to our lives. Yet, when we begin to actually do the applying, the doubting begins to show.

When we believe that Scripture is not sufficient, that belief is manifested across the rainbow of our lives. It covers a spectrum that encompasses a nagging whisper of worry that Scripture just doesn't address all the things we as evolved modern, more intellectual Christians must think about and face, to outright claims that God's Word isn't designed to address all of our needs. Recovering an understanding of the Genesis account as real historical narrative is an important part of a confession of the sufficiency of Scripture. When we claim the Genesis account is not historical narrative, we must understand we have stepped into a view of an insufficient Scripture. Like the church at Laodicea, a view that Genesis is not historical narrative is to be lukewarm regarding the sufficiency of Scripture.

A Biblical Worldview

We must commit to a personal Biblical Worldview. If we don't have one, the first step is to recognize and acknowledge that we don't. This must be rapidly followed by going through a personal inventory of how we view the world and life and comparing our tools of interpretation to Scripture. When we discover our tools for processing truth are not aligned with Scripture, we fully repent by casting away the worldly tools and take up the Scriptural tools. Don't just lay the worldly thinking and processing tools gently down. Through them away.

The Crisis of Understanding

This will require the hard work of study. Study the issues and search God's Word for information on properly thinking about those issues. Do the hard work of building and gaining knowledge. This begins by sitting under excellent teaching in your church. Look for more opportunities to study and grow. Read blog posts from conservative writers who are addressing cultural issues using a sound Biblical worldview. You can also find a trustworthy conservative college and take a course in hermeneutics and Biblical interpretation. If they offer a worldview class, take that.

Become Biblical in Thinking and Acting

- Crisis of Authority
- Corruption of Identity
- Crisis of Understanding
- Confusion of the Gospel

- Confession of Authority
- Clarity in Identity
- Correction of Understanding
- Completion of the Gospel

Genesis is *Real* in History and Space

Corporately, we must do the same. In our churches, Christian schools, Christian organizations – in every part of our community that claims to be bound by the title of "In-Christ", we must be In-Christ. Assess everything you are doing in the light of Scripture. There are many truths that are obvious and need to be applied. That are others that require a mature understanding of a larger compilation of Biblical truth.

A Biblical church will have families that produce families. Young men and women should be giving a priority to marriage and having and raising Godly children. These children should be taught and raised so they give priority to marriage and having and raising Godly children. That's one important part of a Biblical worldview. Maybe a good evidence of its presence or absence in a church. Since God established the family in Genesis 1 and gave the man and woman a command to "fill the earth", that same priority of marriage and children should be present in a community with a vigorous Biblical worldview.

Some parents will need to repent for not insuring their children are properly educated in and with a Biblical worldview.

We must reevaluate education of our children. They should be purposefully inoculated[61] with Biblical truth that includes teaching them the importance of Scripture and the use of truth from Scripture to interpret the world, make decisions, and impact the culture. This responsibility lies with the parents, in a Biblical worldview. Not with the state or the church. This doesn't mean homeschooling is the only option. It does mean that parents are fully informed of the content and methods of what their children are being taught. The parents are exercising control over what their children are taught by the placing them in the right school that have a Biblical worldview.

How Careful Should We Be With The Word of God?

It's reasonable, clear, and ultimate question. How careful should you be with God's Word? Can you be too careful? If you answered, "yes", you have read something else into the question, and most likely made it something that it is not.

---

[61] "indoctrinated" is a proper word for this concept. Unfortunately, it has become a word that produces fear. It shouldn't. To teach doctrine is to indoctrinate. Yes, wrong indoctrination is dangerous. The other side of that statement is that proper indoctrination is healthy. In a culture that worships free-thinking and unbounded knowledge, indoctrination has the connotation of something contrary to being "free" and liberated in doing whatever we want to do. The opposite is the truth. Proper education is liberating. Improper education is enslaving.

The Crisis of Understanding

Not careful enough has led to every error, every failure of the Christian church. Every heresy can be traced back to an individual or a group of individuals not being careful with God's Word. Not confessing that God's Word has the authority and trustworthiness it has. Adam and Eve failed on this issue when they gave into Satan's lie of, "Did God really say that? Did He mean what He said?"

> *Now the serpent was more crafty than any other beast of the field that the LORD God had made. He said to the woman, "Did God actually say, 'You shall not eat of any tree in the garden'?" And the woman said to the serpent, "We may eat of the fruit of the trees in the garden, but God said, 'You shall not eat of the fruit of the tree that is in the midst of the garden, neither shall you touch it, lest you die.'" But the serpent said to the woman, "You will not surely die. For God knows that when you eat of it your eyes will be opened, and you will be like God, knowing good and evil." Genesis 3:1-5*

Claiming God misspoke is a damnable lie. And, anytime we buy into this line of thinking, we place ourselves in danger of damnation.

That's why this is the right time to ask the question. Can we be too careful with God's Word? The answer is an unqualified, unnuanced "no". It is impossible to be too careful with God's Word.

Being not careful[62] with God's Word comes in many forms. One form is a soft carelessness. We can be guilty of soft carelessness by neglect and by purposeful rationalizations. Neglect is no excuse. There are things we should know and do not. Truth from God's Word we should know. Christians have the responsibility to know

---

[62] Yes, I know that "careless" would be used by many others in this case. However, the word choice emphasizes the point. Carelessness implies a lack of action. The use of "not careful" is meant to convey the concept that occurs actively.

the God they serve. We all study what we love. If we claim to love God, then how is it possible to be ignorant of basic important truths about God? This is not an accusation to those who are new in their faith and just beginning their learning. But, for those raised in the church, or who have been, or claim to have been, Christians for many years, to not know some of these basic truths about God is nothing short of purposeful ignorance. Whether a new Christian or an old, when we do not purposefully apply ourselves to learning about God, we are careless. Next to the incarnate Word, Jesus Christ, the written Word is the greatest revelation God has given us.

Most of these purposefully ignorant can quote sport statistics, movie trivia as well as lines from movies, sing along with lyrics from songs that haven't been popular for 20 years. They know all the political pundits and all the pundit's former spouses and children. They have studies what they loved. What they have loved is revealed in their knowledge. When they enter a room where folks are discussing soteriology or ecclesiology or eschatology, they stand quiet and lost. Except for the occasional input based on their feelings.

Start by repenting of purposeful and unintentional ignorance. Recognize the desire to be motivated by feelings to the suppression of thinking as contrary to God's will. The mind in control of feelings is essential to guard against thinking wrongly. To ignore knowledge of God's Word is to be very uncareful with God's Word.

The "hard side" of not careful are those who believe we need to add to God's Word. They believe we must add to what Scripture teaches because culture has changed since the time the Bible was written. Or, they believe knowledge has increased, or we have information not available when the Bible was written. This hard carelessness emphasizes a more intellectual human of today than of yesterday. It is the chronological snobbery first identified by C.S. Lewis.[63] You will run into this hard carelessness in the hard sciences such as biology, physics, geology, and astronomy. However, it also is dominant in soft sciences like sociology and psychology. Experts and non-experts in each of these disciplines argue they add information not in Scripture that enhances our understanding of the underlying issues that Scripture tried to address. And, sometimes, corrects what Scripture addressed at the time written, but is not applicable today.

Examples of this type of hard carelessness include the issue of women in the roles of pastoring and preaching God's Word. Even in the face of very clear teaching, the current battles over egalitarianism vs complementarianism of the sexes is a battle between hard carelessness and hard carefulness.

It comes down to this: Is God's Word sufficient?

---

[63] Although others may have identified this issue in other ways, it is generally attributed to C. S. Lewis in relating a conversation with another member of the Inklings, "Barfield never made me an Anthroposophist, but his counterattacks destroyed forever two elements in my own thought. In the first place he made short work of what I have called my "chronological snobbery," the uncritical acceptance of the intellectual climate common to our own age and the assumption that whatever has gone out of date is on that account discredited." *Surprised by Joy, The Shape of My Early Life,* ebook based on edition by Geoffrey Bles, London, 1955, p 143. Project Gutenberg Canada, Ebook #1275.

ABOUT THE AUTHOR

Dr. Marks is a Professor of Chemistry at a Christian University in South Carolina. He has served as an Elder, Music Leader, and Sunday School teacher in local churches. Before teaching Chemistry, Dr. Marks spent twenty years in the United States Air Force, where he served managing the acquisition of aircraft simulators, leading and participating in chemical and biological intelligence work, and teaching at the United States Air Force Academy. Dr. Marks earned three academic degrees from the University of Tennessee at Knoxville: Bachelor of Arts (Chemistry with Honors), Master of Science (Chemistry), and Doctor of Philosophy (Chemistry). He is a member of the Creation Research Society and the Military Officers Association of America. His research interests include the study of creation, the relationship of science and Christianity, polymer supported organometallic reagents, and computational chemistry.

Made in the USA
Lexington, KY
30 October 2019